THE INVISIBLE ANATOMY

THE INVISIBLE ANATOMY

DISCOVER THE INTUITION OF THE HUMAN BODY

J.K. DICKINSON

Expression Through Words, Inc

My mother, Judy, filled my heart with unconditional love. Her sensible thinking and 'never give up' attitude gave me my internal strength. I love you mom, G

Copyright 2016. All rights reserved.

Intellectual property contained herein protected domestically and internationally by patent, trademark, and copyright laws. No part of this book may be used or reproduced by any means, graphic, electronic, or mechanical, including photocopying, recording, taping, or by any information storage retrieval system without the written permission of the author except in the case of brief quotations embodied in critical articles and reviews.

ISBN: 978-1-7336805-0-9 (Hardback edition)
ISBN: 978-1-7328495-2-5 (Paperback edition)
ISBN: 978-1-7329495-0-1 (ebook / Epub edition)
ISBN: 978-1-7328495-1-8 (ebook / Kindle edition)

1st Edition 2019

Publisher: Expression Through Words, Inc

Photo credits: Cover: Shutterstock © Dan Howell; background NASA image, gold borders Shutterstock © Hakinmhan. Image modifications by Michael Rohani. Book interior pages: Page vi, AdobeStock © MozZz; pages 3, 7, 9, 11, 14, 17, AdobeStock © Axel Kock with modifications by Michael Rohani; page 17, Shutterstock © Giovanni Cancemi; page 19, iStockphoto © Xrender; page 20, Shutterstock © Antonio Guillem; page 21, iStockphoto © Xrender; page 29, iStockphoto © Sudok1; page 31, AdobeStock © Giovanni Cancemi; page 34, iStockphoto © Solvod; page 36, Energy of Life diagram by Michael Rohani, female figure, iStockphoto © Janulla, background image from NASA, gold borders Shutterstock © Hakinmhan; page 40, Shutterstock © Grischa Georgiew; page 46, AdobeStock © Arseniy-Krasnevsky; page 53, AdobeStock © Iakov Kalinin; page 62, AdobeStock © Blair Costelloe; page 71, AdobeStock © Juliet Photography; page 76, AdobeStock © Lorenzo Patoia; page 81, AdobeStock © Masson; page 84, AdobeStock © Christian; page 86, Shutterstock © Danilov1991xxx; page 90, AdobeStock © AboutLife; page 93, AdobeStock © Chlorophylle; page 94, AdobeStock © Giovanni Cancemi; page 95, iStockphoto © Grafissimo; page 96, iStockphoto © Natalie Board; page 97, iStockphoto © MonkeyBusinessImages; page 98, iStockphoto © M-image; page 118, Shutterstock © Kateryna Kon; page 124, AdobeStock © Jozefklopacka. Authorized client photos and personal archive photos: pages 100, 101, 103, 105, 107, 108, 110, 111, 112, 113, 115, 127. Page 131 and backcover, author photos by Lis Larkin.

Edited by Mikel Benton
Book design by Michael Rohani | DesignForBooks

Printed in the USA.

DISCLAIMER:

The information contained herein is not intended or implied to serve as medical or therapeutic advice, diagnosis or treatment. You should not use this information to diagnose or treat any health issues, illnesses or diseases. The book is not providing any medical or therapeutic advice. The information contained within is based on the professional experiences of JK Dickinson. If you have or suspect you have a health or medical problem, you should consult with your medical provider or professionals.

CONTENTS

Introduction   ix

Foreword by Huy Hoang, MD   xiii

CHAPTER 1
The Invisible Anatomy—The Neurons   1

CHAPTER 2
The Energy of Life System   37

CHAPTER 3
Three Energies That Destroy Hope, Happiness, and Health   85

Illusional Energy   85
Expectation Energy   90
Story Energy   93

CHAPTER 4
Courageous Individuals—Hitting Life Head On   99

EPILOGUE
Know Your Truth   121
Traits of One Nobel Laureate   127
Special Profile in Courage   129
Acknowledgments   131
About the Author   133
Books by J.K. Dickinson   135

> THE CREATOR of this incredibly sophisticated human body did not forget to provide us with access to infinite intelligence and the doorway to knowing the power of the human spirit!
>
> J.K. DICKINSON

INTRODUCTION

Our world, the world we are currently occupying, has been designed with impeccable care and meticulous creativity. The intelligence is far superior to anything we can possibly imagine, even for those of us with extraordinary imaginations. The next time you have a free moment, head to the zoo or an aquarium and take a good hard look at the animals, creatures, and mammals; take in their energy, their design, their sensing mechanism. We know that animals have extraordinary sensing ability. Imagine now what bells and whistles the human body has. This world, the planet earth, is a creation that most of us will never understand in its entirety while we are alive. The design is too massive. The physical body we move around with has an extremely sophisticated wiring system. It has a sensing system, an intuitive ability that is beyond magnificent. I intend to introduce you to the sensitivity and power that live and are available to you within the human physical body.

Many of us take our senses for granted. If we pause for a moment to consider nature, the entirety of the planet's habitats, you will recognize that animals, for instance, sense danger, the location of food, and the direction to travel without being taught. Learning about the sophisticated intelligence living within yourself will allow you to become more confident. Learning to harness its power and relying on it will change your life.

This entire planet is strategically designed to allow us to live, thrive, and enjoy the moments of existence. We see it time and time again in the plant world, the animal world, the incredible ocean, and so much more. For a moment, please take into consideration that the planet we reside on rotates twenty-four hours a day in precise movement, allowing our entirety to live. Twenty-four hours a day, three hundred sixty-five days a year, without any interruption and, quite frankly, without anyone at the helm.

If you take a moment to get in touch with your senses and feelings, you will discover the power you hold within your reach, a tool that will serve you a lifetime in ways you may find unbelievable. We've heard of the miracles that take place in healing, our strength that represents itself during times of fear, and other unexplained phenomena. The strategic sensitivity and unlimited power that lives within the human anatomy is genuinely extraordinary. Let's now begin the journey to understand this unknown vehicle in which your physical body operates. We will incorporate the connection to your spirit energy, spiritual strength, and intuitive intelligence.

I believe that our spirit, the energy or essence of who we are inside this physical body, is the energy we arrive with when we undertake this life journey. We are a dominant species with unimaginable power and capabilities.

My name is J.K. Dickinson. I specialize in reading energy associated with imbalances within the physical, mental, emotional, and spiritual components of the human body and psyche. With my natural gift of intuition, I can swiftly and accurately detect challenges that cause various types of discomfort, some leading to serious physical illness.

For the past twenty-five years, I have dedicated myself to helping others by carefully and passionately creating an infrastructure of vital information that can and has assisted private individuals, organizations, scientific research teams, and licensed healthcare practitioners. It is through my accurate understanding of the energy attached to dysfunction, illness, stagnation, and unhealthy thoughts and feelings that I can open the door to hope, answers, and factual data.

Fascinated with the stream of energy relating to illnesses and long-term imbalances, I began to study patterns associated with individuals who requested my assistance. I developed charts to understand where, when, and how long the experience or memory had affected the person's well-being. I also traced the pathway that the neuron cells traveled while carrying the energy and stored data. The journey for me has been remarkably fascinating. In 1996, when I aggressively embarked on understanding more, I had no way of knowing the wealth of information I would gather along the road, twenty-plus years later. In 1998, I created the Energy of Life System. My basic understanding of the energy pockets strategically placed to alert the physical body of weakness, illness, and imbalances were eye-opening. This system, the Energy of Life System, has been used by licensed health providers for almost twenty years.

Today, I am excited to introduce my new book, *The Invisible Anatomy*. In 2009, I created the Invisible Anatomy System. Over the past nine years, the Invisible Anatomy System has gained credibility as a tool to focus on the vagus nerve and the neuron activity that causes severe physiological, neurological, and human musculoskeletal system damage. I intend to provide valuable insight to those who want to understand how life experiences can affect specific areas of the physical, emotional, spiritual, and mental energy that runs through the body, via the vagus nerve. Moreover, you will find insight into the neurons, the live cells, and how they affect our bodies, rendering a balanced self or an imbalanced self.

You will discover how the energy that runs through your physical body via the neurons will allow your body to remain balanced and energized with easily accessible power or, on the other side, depleted energy resources, anxiety, depression, nervous system challenges, and so much more.

I am a person who loves nature and music. The vibration and energy of each have an immediate effect on the nervous system, brain, and stomach. Music and nature can instantaneously affect the energy of the human mind and body. Natural outdoor resources are one of the strongest and quickest ways to ground the emotional and mental chatter: the ocean, lakes, and mountains are all packed with energy to ground the physical, emotional, and psychological areas of our beings. In my alone time, while I am in nature, I often wonder how the planet, its inhabitants, and the "whole" thing works in harmony without anyone at the helm. And yet, the creator of this entire experience has left no stone unturned. Music integrates the vibration of the notes, instruments, and words to make a powerful energizer or relaxer to the human body and nervous system.

As we all know, there is only one guarantee: we will live for a moment and we will pass away, in a moment. In between these two events, life moments will arrive when the time is right for our growth, without any planning. I hope that within these pages you will discover a few tools that will allow you the wisdom to unite with the unlimited power you hold within yourself so that your life journey includes your version of happiness, your connecting to your dreams, and hopes, and so on.

If you take the time to look at every pocket of life on this planet, including animals, mammals, sea creatures, plants, water channels, land masses, and much more, you will know the level of untapped intelligence available to us—the planet we live on "works" naturally.

As you being to read *The Invisible Anatomy*, I encourage you to pause along the way. Learning about real-life stories that may be similar to your own can provide hope, acknowledgment, and a connection to a potentially new plan of attack against current challenges in your own life. Let me also encourage you to listen to your own body's reaction to what you are discovering on the pages. There is no more significant time to connect to and get to know the energy that is pulsating through your spirit and your body than now.

Let's now begin with the **energy** of this incredible system within the human body, **the Invisible Anatomy**!

FOREWORD

HUY HOANG, MD

After years of study, my current understanding of the human phenomenon is that it is composed of three interconnecting parts: the mind, the energy body, and the physical body. Of the three parts, the energy body is the least studied by science. At this time, the energy body, such as the chakras and acupuncture meridians, cannot be proven by Western medicine. There are attempts to prove the existence of chakras such as the Apparatus for measuring the functioning of the Meridians and their corresponding Internal organs (AMI) by Dr. Hiroshi Motoyama, the Gas Discharge Visualization (GDV) by Dr. Konstantin Korotkov, and various studies to delineate acupuncture meridians, but they cannot be considered definitive proof. I know it exists because, in my clinical experience, I have witnessed the improvement in the health of my patients when they go through a balancing process of this energy phenomenon.

The energy body connects the mind and the physical body. Without the energy body, the mind cannot direct the physical body to action. Without the energy body, the physical body is just a lifeless body, like an inanimate object.

Unlike other books filled with mystical Eastern terms, this book gives down-to-earth descriptions of the energy centers, the vagus nerve, and the critical role the neurons play in the health of the human body. Instead of preaching theories, this book describes practical examples with actual cases. The reader can see that when faced with life events, the mind reacts to give rise to imbalances in the energy centers leading to physical illnesses. These cases confirm what I have encountered through clinical experience with patients.

In addition to the information about the energy centers that Jennifer has explained in her previous books, there is now information on the invisible anatomy and the vagus nerve. The vagus nerve is a brain nerve called the 10th cranial nerve which has connections with many organs, such as the lungs, heart, stomach, small and large intestines, liver, pancreas, spleen, kidneys, and more. Since it has a significant role in the parasympathetic nervous system activation, it connects stress, the autonomic nervous system, and those physical organs. Therefore, it plays a major role in mind-body medicine in general, and the gut-brain axis in particular. Currently, Western medicine is exploring vagus nerve stimulation for various diseases such as seizures and depression. I believe that in the future, Western medicine will confirm the role of the vagus nerve in other conditions as described by Jennifer.

I invite you to read this book. You may recognize some imbalanced energy patterns described in this book in your own life. If you correct these imbalances, the first area of empowerment is your knowledge. You may find that your imbalances and illnesses will improve.

Best of Health,
Huy Hoang, MD
Natural Health Medical Center, Los Angeles

Dr. Huy Hoang began his academic journey at the Massachusetts Institute of Technology in 1984. Discovering his interest in medicine, and the possibility of treating and fixing health challenges, he graduated with his medical degree in 1988. From 1988 to 2009, Dr. Hoang worked in internal medicine at prestigious facilities in Los Angeles. In 2009, he established the Natural Health Medical Center in Lawndale, California. Realizing that both traditional medicine and alternative therapies share validity in patient care, Dr. Hoang opened his doors.

THE INVISIBLE ANATOMY
The Neurons

The Invisible Anatomy has Five Components:

THE VAGUS NERVE, EYES, BRAIN, STOMACH, AND HEART

I'M EXCITED to introduce you to information that I feel is invaluable if your goal is to take your health and wellness into your own hands as an active participant. One of the most important points I would like you to take away from this book is the power the neurons and vagus nerve have while traveling through the physical body. You will discover that there are billions of these cells moving around just waiting for you to give instructions. The intricate electrical system that is working for or against you inside the human body is not only impressive, but it's also simple to understand if you keep an open mind. The creator of this human body equipped us with everything we need to maintain balance, attract our hopes and dreams, and live with unlimited energy.

To uncover what this human body is capable of and equipped with is an extraordinary undertaking. Once discovered, you will never want to put this information away.

Based on the past two decades of my work, I am confident to introduce the concept that the eyes, brain, heart, vagus nerve, and stomach feed energy into the physical body, with the neurons acting as the transportation agent. My area of expertise allows me to pinpoint events, moments, and circumstances that create blockages in critical areas of the physical body that eventually form into diseases.

As I worked to determine a conclusion about diseases and potentially fatal illnesses by identifying the nervous-system pathways that were instrumental in disrupting the flow of energy necessary to live in full constitution physically, emotionally, mentally, and spiritually, I discovered the energy centers. That was the first piece of the puzzle for me, in 1994. The patterning I saw with my gift of intuition directly linked the locations attached to various illnesses, whether in the physical body, emotional body, mental body, or spiritual body, to the energy centers.

In this book, *The Invisible Anatomy*, I will refer to the five areas in which illnesses and imbalances can grow, as they all have their energies and singular energetic components when it comes to illness or imbalance.

Energy and how it interacts with us on a daily basis is the foundation of my work and teachings. The energetic component of all words, thoughts, hopes, dreams, intentions, expectations, fears, anxieties, and life movement has power.

While reviewing the neuron activity in the five areas of our discovery, we will pay very close attention to the vagus nerve. The vagus nerve is the longest nerve in the human body, and it connects the brain to organs, the heart, the stomach, and so much more. The vagus nerve has not been a focus until recently. As these discoveries involving the vagus nerve are in research development, the medical and science communities will surely expedite their focus. The health and constitution of this nerve are imperative to the overall health of the human body.

VAGUS NERVE FUNCTION

THE INTUITIVE NERVE CONTROLLING MAJOR BODY FUNCTIONS

The most powerful nerve in the human body:

- Heart rate, cardiovascular health
- Respiratory, breathing
- Digestion, gastrointestinal health
- Throat function and speech
- Muscle health
- Brain function

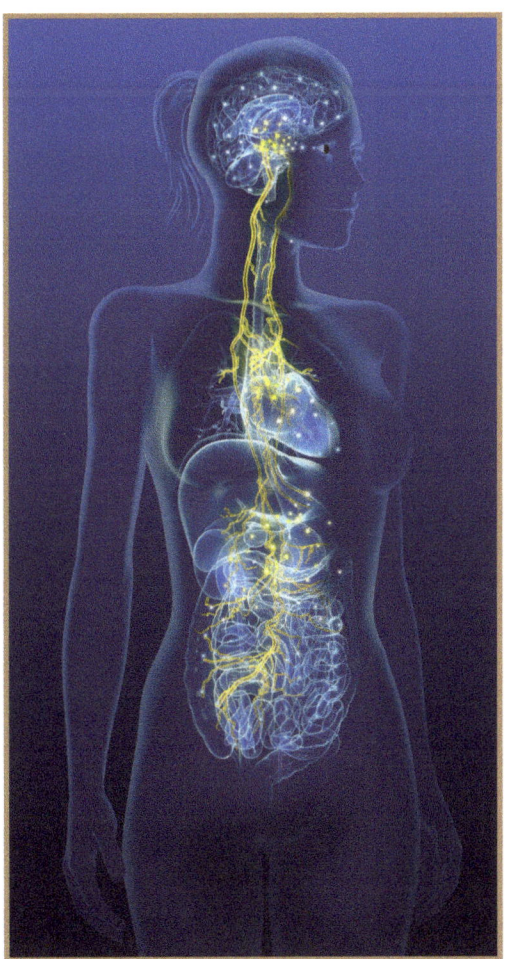

Unprecedented Impact

To date there is very little data outlining the magnitude of the impact the vagus nerve has on the human body and human spirit. This nerve is the most potent nerve in the human body and must be understood if medicine and science will gain the edge on the crippling challenges associated with and from this incredible energy source. I believe the vagus nerve is the culprit in significant health challenges of the twentieth century that include MS, bipolar disorder, Parkinson's, Alzheimer's, dementia, anorexia nervosa, Crohn's disease, and SARS, to name a few.

I've been charting client and patient diseases, imbalances, and nervous system-related disorders for over twenty years. In my opinion, MS, Parkinson's, Crohn's disease, anorexia, anxiety disorders, IBS, insomnia disorders, bipolar disorder, clinical depression,

ADHD, ADD, and obsessive-compulsive disorder all begin the journey from healthy to unhealthy with a malfunction associated with the vagus nerve. Although I am not a licensed medical professional, nor a physician, I am qualified to share my experiences over the past two decades. The vagus nerve is responsible for the malfunction of nerves, neurons, muscle tissue, brain waves, digestion, balanced hormones, and so much more. A healthy vagus nerve allows many of us to live with a lively, non-diseased physical body and mentally/emotionally balanced thoughts. An unhealthy vagus nerve, well, imagine what havoc can take place with the organs, brain, muscle tissue, nerve channels, and more.

As you read through these pages, I am sure you will find that you have already known, in some form or fashion, the truth of it all. Your stomach has given you signals. Your heart has told you things that only you would know to be true. Moreover, you've had instincts or a "knowing" from an early age. I'm sure you can look back for the "aha" moments of realization at different times of your adult life as well.

The Invisible Anatomy is a sophisticated electrical system aligned with the highest level of intuitive knowledge known to date. It works twenty-four hours a day without any prompting, commands, or instruction. Learning about it, following up with it, and practicing a daily awareness of it until it becomes a natural pattern within your lifelong toolbox will allow you to be the leader in your first line of defense, supporting your health, wellness, dreams, and goals. To live each waking day with the key to igniting and using this indescribable intelligence has immeasurable advantages. You can benefit in ways that include understanding pattern challenges both emotionally and mentally, connecting to your intuition and wisdom from birth, and finally, discovering an easy-flowing level of self-confidence that will last your entire lifetime.

Let's take a moment to address the natural phenomenon of intuition. Intuition is a word linked to infinite intelligence, the most sophisticated intelligence available to a human being. Today, more than ever, intuition and any association with the word are in demand. Those who require the edge in their businesses, their careers, their health, and their conscious awareness have found ways to connect to their intuition and grow with it. Intuition is a natural tool available to those who open their hearts, feel the signals through their stomach areas, shut down their minds for a moment, and explore their senses. It arrives with us at birth. Some want to acknowledge it and learn more, and there are those who choose to keep the door closed. Millions of people spend billions of dollars on self-discovery and self-help to learn as much as they can about health, relationships, life, wellness, consciousness, joy, success, abundance, and living longer! Billions of people around the globe want to live better, longer, and with a great deal of happiness and wealth. Why not? It's certainly available to each of us if we believe. At least that's what is being communicated to us, over and over again.

Everywhere I look I see the words "intuition" and "intuitive." More and more influential people have announced their connections to intuition. I listened to an ad for a master class the other day in which award-winning director Ron Howard mentioned his intuition. Steve Jobs was a big promoter of intuition as well. Intuition is a natural awareness, an intellect that exceeds normal comprehension. That's it. Our society is no longer burning witches, associating intuition with palm readers and so on. The desire to develop intuition is multiplying around the world. A large percentage of the most successful people in the world have mentioned their knowing, gut feelings, or senses in their remembrances of their successes and overcoming hardship.

To take your health and wellness into your own hands seems to be a mandatory requirement in today's world. More than ever, human beings feel excessive levels of stress, anxiety, and overwhelming levels of pressure. Four out of ten people in the United States take anti-anxiety, anti-depression, blood pressure, or hypertension medication. Sadly, children under the age of ten are prescribed medicines for symptoms that twenty years ago would not have been the norm; it would not have been acceptable in our society. However, with stressed out parents, the need to medicate families is seemingly regular these days. In my work, I am not able to judge the epidemic with medicines and prescriptions as each person and each family, have the situations they are dealing with within their environments. With over sixty billion dollars spent on self-help and prescriptions, the search for more in-depth answers and resolutions speaks for itself.

Learning how to navigate the Invisible Anatomy with an understanding of the Energy of Life System will allow you to have a deeper understanding of illnesses and the feeling of imbalances. I know that after reading a few chapters, many of you will start to remember how many times in your own life you felt something and knew it was the truth; you knew it to be right. It can be a confirmation for those who have always known they had connections to their intuition and the pulses of their senses.

The individuals, families, groups, and professionals who have contacted me for support have one thing in common: they've searched for answers to questions that year after year went unanswered, or the "advice" wore off without lasting results.

You will find stories throughout this book from individuals and families who have overcome life challenges to emerge with a "knowing" that the energy that surrounded them is now understood! It will give real-life evidence that you too can make shifts and forward movement in your own life.

THE EYE

The Eyes See It All

OUR EYES HAVE PHOTORECEPTORS THAT PHOTOGRAPH LIFE

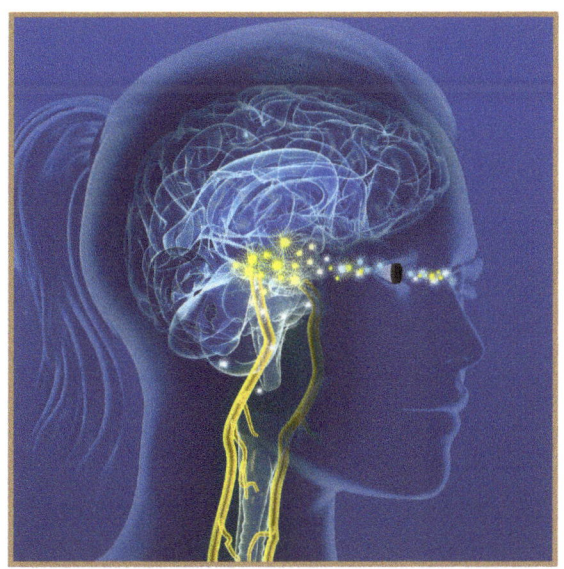

There are over 1.5 million retinal ganglion cells in the human retina. Many things you have visually engaged in have been stored, have traveled, and can affect you either positively or negatively, without notice.

Imagine the life events that are joyful, exciting, filled with love, and quite memorable. You will want to cherish these moments and hold on to them for future review and energy. We are often driven by memories that gave us hope as well as experiences of feeling and being loved.

Let's turn this around now and switch it up. How about memories filled with pain, unimaginable abuse, unexpected visual encounters, and feelings of being unloved, unworthy, and detached from the warmth of family.

What happens when a new moment triggers that mental photograph, the painful memory, or an unexpected circumstance? For many people, it's a regular part of an unhealthy environment they continue to visit. Medication is often needed when the stress, anxiety, and fear set in.

For others, the human spirit will kick into gear to create a plan of escape—a plan to remove itself from any further destruction, pain, or fear. When that door of hope and possibility opens, through effort and dedication, miraculously "aid" will arrive in many forms. The energy the spirit used to ward away danger, pain, and despair wins! Hearing stories of survival, success against all the odds can bring us a feeling of excitement, a connection to hope, and an "I can do it too!" mentality. When we watch or read individuals discussing how they were able to overcome similar challenges, the body begins to gain energy and hope, and the neurons begin their work. New photos, new images of hope, and the voltage is set in motion to diminish the earlier stagnant, unhealthy energy.

We champion the winners of the most profound life struggles and somehow believe and intuitively know that it's possible for us too! Regardless of what you have seen in your life, if you are breathing and want to engage in your life, you have an opportunity to bring in new photographs, new awareness, and the ever-so-imperative positive mental words.

Intuitive Connection

SENSORS CONNECTING TO INTANGIBLE AND TANGIBLE TRUTHS

Your eyes can be used as sensors allowing you to see the truth. It can be as subtle as an expression on a person's face, focusing on an object that sends a sensation to you, or discerning when a person's words are not in balance with the context being spoken. Using your intuition, when you look back at a moment or event, you are able to discern the truth of the situation through your feelings, led by the images stored through your eyes. The eyes are able to see outside the tangible dimensions currently known to mankind.

THE BRAIN

The Mega Data Machine

RIGHT—WRONG, INTELLIGENCE—IGNORANCE, ACCURATE—INACCURATE

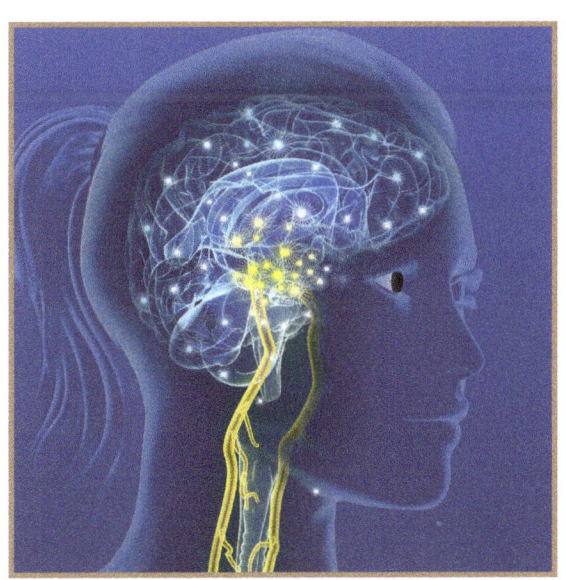

The brain receives data, stores data and sends data. That's it. It controls the thoughts, memory, speech, muscle movement, and many of the organ functions. The brain's role in connection to the nervous system is dependent on the neurons. Sensory neurons access information from the eyes, ears, nose, skin, and tongue and send it to the brain. The brain will then send the news throughout the physical body. Imagine all this going on without your conscious participation or, for the most part, realization. For most of us, without the connection to the sciences and medicine, we could be clueless as to how the entirety works together.

The brain receives information from all areas of the body. With this information, a signal is sent to give instructions. It's an automatic response. Dependable or not, it will provide you with messages and stored thoughts. The brain does not need to think whether the reaction is right or wrong; it gives saved information and instruction. Although a memory, a feeling from the past, is history, even the most straightforward implication of the sequence of events that lead to the memory or thought can trigger a response from the past. Remember, the brain stored the information. The brain does not keep a calendar or a memo confirming that you have moved past the last moment unless you have filled up the data bank with *new* information to override that experience or moment.

Pausing for a moment to reread this paragraph can be helpful for those trying to change patterns in their lives. The brain is fed information, right or wrong, truthful or illusional; it sits and awaits information, signals, thoughts. The brain is being programmed from birth, establishing intentional and unintentional patterns.

Feed your brain with new information each day. Feed it hope, joy, acceptance, memories of good times, and feelings of love, allowing this "food" to change the neuron energy that circulates in your body! Why let your neurons continue to eat, digest, and transport throughout the body the same old stuff if you can create new energy? By creating new thoughts, empowering your heart with meaningful feelings and emotions, you will increase your energy levels, reduce the anxiety and fear, and, most importantly, allow to live each day with possibilities!

Who in the world today does not need or want extra energy?

The best news is that your eyes are waiting to photograph all your dreams, heart's desires, and moments you want to remember, *by sending the data* to the brain.

Intuition Connection

INFINITE KNOWLEDGE PORTAL

The brain has an area that allows a sensor to accept or reject information if the participant is aware of the signals and sensations. Although the brain stores information, there are three specific areas of the brain that can receive stimulation that opens the channel to infinite knowledge. This area is not in the data storage areas. Many of the successful men and woman who have high-profile responsibilities in life learn to use this area of their brains with ease.

THE HEART

The Intuitive Feeling Center

THE MUSCLE WITH A BRAIN

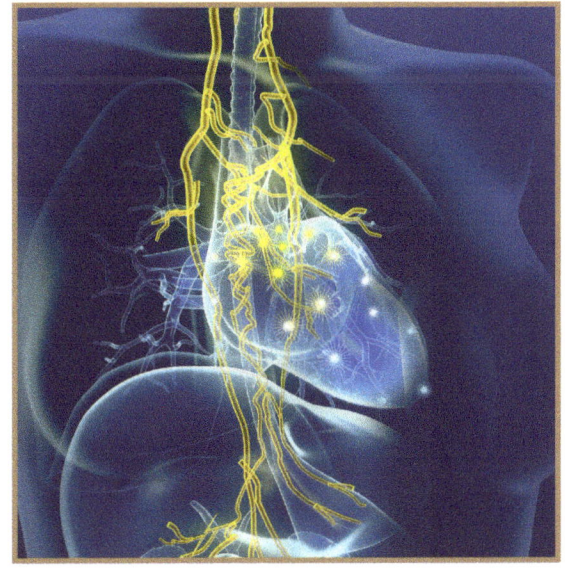

From an early age, we begin to feel life. Without expression of words, without an understanding of words, we sense life. Our heart helps us to understand what our spirit knows to be true. We hear voices, words, and sounds, and we respond to that stimulation, through our feelings.

The human heart has a similar structure to the human brain. With the heart's nervous system, it sends and receives signals twenty-four hours a day, every day of our human lives. The heart endures far more than we realize when we do not listen to the messages. Blood pressure can go up, or quickly descend. The heart center can send signals to the stomach, giving sensations of nausea, dizziness, and so much more, to get our attention. The blood circulating through the heart can become invaded with various impurities if the heart pressure is not relieved. The energy from the heart is transported via the neurons, the vagus nerve, and the blood. Somewhere inside us all, we know that the truth is at the "heart" of all human matters. Our heart is in constant contact with the brain. It recognizes the truth, has a direct link to intuitive intelligence, and will guide us with the energy necessary to engage in life to the fullest capacity possible, regardless of our past or current situations.

If the stimulation in our environment is soft, caring, and protective, our muscles will remain relaxed and our nervous systems will remain calm. Our organs, vagus nerves, muscle tissue, and brains will not become overstimulated, overwhelmed, or fearful. The physiology of our bodies will have the time necessary to grow within the mandatory timeline for growth without conflict.

On the other hand, if an environment is filled with anxiety, anger, and lack of touch or soft voices and nurturing, the muscle groups, nervous system, and vagus nerve will begin the journey of imbalances and illness.

I have discovered that the spirit that lives in the physical body will, at all costs, try to save and protect the heart and the feelings associated with it. The heart knows. Love energy is one of the most potent, natural, intuitive areas of the human being. Given a strong understanding of heart health and heart detriment, a person leading with his heart allows himself to enjoy his life thoroughly. It's a twofold success if you are open to give and receive love. It's beautiful for the person giving love as well as those fortunate enough to encounter the experience of being loved.

One emotion, one feeling, that will have a direct effect on the heart is the word "love." The word love is sacred for many people. Please be mindful of its implications when using the word. It's a word not easily communicated by some, and when it is, there is almost always a meaning behind it. Unless you've experienced this understanding, the word can be flung around like a scarf around the neck. Love should not be a word for sale if the heart is connected. Love is a feeling, a gesture, a symbol of care and support.

One of the strongest energetic impulses to a human being is actually to feel loved by someone. It can shift long-embedded energy into a sturdy, impulsive action rod at any moment of a person's life. Love has unmeasurable power!

What happens when you "feel" something that may be true and real and you override that feeling or knowing, most often with data from the brain? Individuals who have a connection to the intuition of the heart have assistance any time during any moment of the day.

Give the heart a chance to help you with decisions. Living with an open heart filled with love and acceptance can bring wonderful gifts and experiences your way. Your heart has many answers if you will allow yourself to listen.

Intuition Connection

FEELING THE TRUTH OF EMOTIONAL INTELLIGENCE

The heart intuition senses emotion, clarity, and truth attached to data, feelings, touch, and sensory input. The heart has access to infinite knowledge, based on the individual's human spirit. Each human soul has a particular purpose and unique energy. The human spirit emanates the heart energy. The heart of each human being is ready to support a balanced life filled with joy, happiness, calm, and realized hopes and dreams. Heart energy is attached to the human beings' purpose and desired experiences in this lifetime.

THE STOMACH

The Gut Feeling

THE VAGUS NERVE: THE VAGUS NERVE CHANNEL IN THE STOMACH AREA IS ONE OF THE MOST POWERFUL SOURCES OF INTUITION IN THE HUMAN BODY.

The stomach sends signals to the heart for support. The human emotions are directly connected energetically to the stomach, which

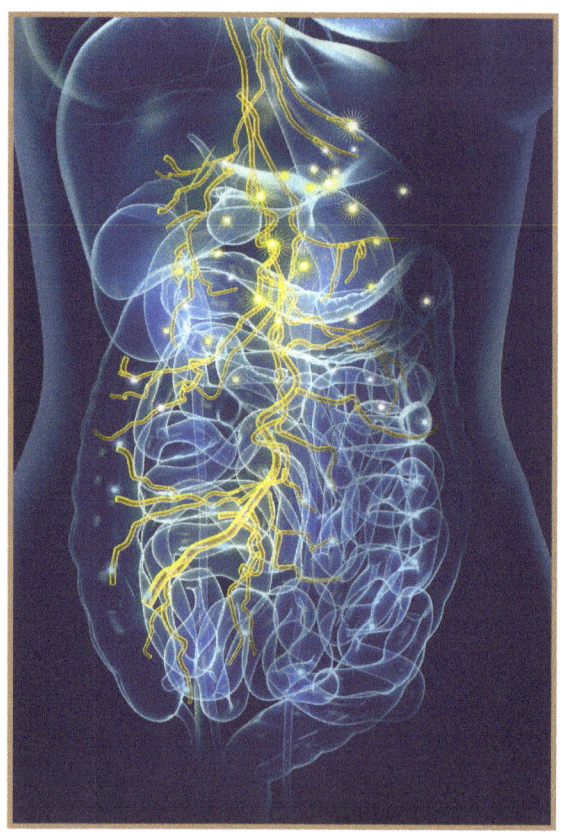

sends signals swiftly and intensely to the heart and the nervous system.

The stomach area of the human body has millions of neurons sending information, processing information, and storing information instantaneously, while we are awake and while we sleep. The stomach neurons send energy to the heart, vagus nerve, and the nervous system. It is the area that, when compromised, can begin the journey to the following diagnoses: MS, Crohn's, IBS, anxiety disorders, anorexia, chronic depression, ADD, ADHD, digestive disorders, and more. The stomach area is energetically attached to self-confidence, self-assurance, how we are seen and felt in our environment, and so on. Our stomach, the abdomen, governed by the vagus nerve, is such a critical area of the human physiology.

The nervous system is the main control panel of the human body. Voluntary and involuntary impulses and signals come from the nervous system. I want to focus on one area of the nervous system, the vagus nerve. This intelligent, significantly important nerve is the longest in the nervous system family. The electrical wiring running through the vagus nerve is responsible for the health of the digestive tract, heart rate, respiratory system—lungs, major organs, and muscles. It is the nerve that connects the brain to the physical body.

While tracking client insomnia challenges, I discovered the reason for sleeping pattern disruption in one group. The events that caused shock, fear, disbelief and confusion, alerted the neurons to become overexcited. The neuron track was broken sending irregular electricity to the stomach, heart and brain. Each had similar problems just before experiencing insomnia, life events that caused fear and a disconnect from a loss. The repetition of the thoughts and memories traveling through the nervous system of these individuals did not allow for restful nights. The food of mental chatter traveling through the stomach to the brain was causing a series of breaks within the neuron channels. It is akin to jumping aboard an out-of-control train carrying toxic waste up a very steep hill.

With every new piece of the story circulating—the painful past, the photographs in the mental mind—the more emotions began to add in: anger, timelines for accuracy and confirmation, what if's and, eventually, attempts to glue a more comfortable picture together while the train was moving full speed ahead.

The result: full-blown panic attacks, inability to breathe with ease, headaches, and nervousness resulting in shaking. So in addition to insomnia, they introduced anxiety, shortness of breath, and an overall feeling of sickness.

In individuals who also suffered from anxiety and fear, once the nervous system caught wind of thought, that would repeat the pattern of driving the train of insanity, causing the entire body to begin to shut down. The transmissions we send through our minds to the nervous system via the neurons can be dangerous to our mental, emotional, spiritual, and physical health. Discovering how our thoughts and experiences affect our physical bodies is imperative.

The nervous system sends immediate signals via the neurons to various parts of the body throughout the day. In balanced individuals, the energy surge is joyful, exhilarating, and filled with excitement and endorphins. It's particularly true when there is love, passion, or creative energy involved somehow. When a person lives with calm, peacefulness, and, more importantly, satisfaction, the physical body, the mental and emotional component, and the spirit are at ease. It's a perfect cocktail for a very happy ride to the destination created by the thought. The digestive tract will absorb nutrients without significant discomfort, and life will be balanced.

In the individual who is imbalanced, the ride is entirely different. The memories and thoughts will send tons and tons of flooding neurons filled with the energy of fear, doubt, anxiety, and confusion through the entire nervous system, muscle groups, heart area, and, finally, the brain. For the person with painful, angry, or confused thoughts or visions, these moments can be debilitating. The person may have an issue with absorbing nutrients from her food choices, leaving her without the protein necessary to stabilize her thoughts, perhaps the inability to concentrate for a considerable amount of time, and with a short temper due to the restless feeling inside. It's a vicious cycle for some.

I know that the forty million Americans who suffer from anxiety-related challenges, and the one out of every six in the United States who are on anti-anxiety medications, do not understand the "train ride" theory above. They do not know that it's possible to breathe through the destructive thoughts before they hop on the train of emotional, mental, and physical destruction.

Intuition Connection

THE HUMAN SPIRIT'S UNLIMITED POWER SOURCE

The stomach area, filled with millions of neurons, is the center of the power of intellect sensory which aligns meticulously with the area of the brain connected to personal wisdom. The center of the human body will lead a spirit into calm, happiness, realized dreams, and much more. This area will also drive the physical body into turmoil, despair, anxiety and debilitating diseases, taking important human life experiences away from the spirit. The stomach is a vital area of inherent power.

THE MEMORY TAPE—
STORED INFORMATION

Neurons Are Information Messengers Using Electrical Impulses and Chemical Signals to Transmit Information

Imagine a storage tape filled with significant events that have occurred in a person's life. I am familiar with this phenomenon because I have access to this information through my gift of intuition. It may be hard to imagine or, quite frankly, believe for some, and yet, it is the truth. When I see the events "stored" in various energy centers of those I have spoken to over the years, as you can imagine, it's at times startling. Year after year, my fascination grew. I wanted to know as much as I could about this astonishing system of storing.

Before undertaking this work, I was not a true believer of hypnosis. It just seemed a bit farfetched to me. I understand the theory of hypnosis now and know why it is a useful method within the mental health industry, especially with a qualified practitioner with the expertise of stored data.

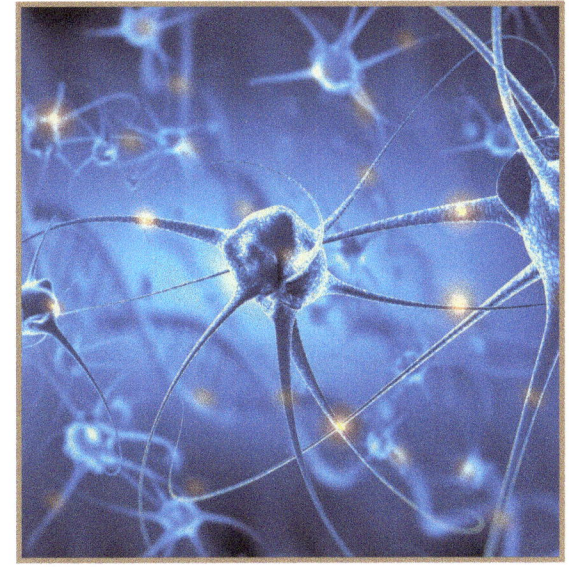

In 1998, my long-time approach to case study research was turned upside down when I was asked to review a case of an individual who had just been diagnosed with bipolar disorder. This person had not slept in several months, could not hold down food on a consistent basis, and had become very anxious. While watching the "film" of the person's life through my gift of intuition, I noticed three life events that had created the energy line of disconnection. I recognized three key moments endured by this young man that had left his emotional state in an extremely fragile place, as his brain tried to process, accept, and layaway "with any stability" the data.

Imagine watching the tapes of what transpired in a person's life that had left him without calm control of his mental mind and his emotional self. I watched with amazement how this information affected the entire psyche of his whole being. For several years afterward, I became quite interested in studying life events amongst individuals diagnosed as bipolar. Because I can access the memory tape quickly and with accuracy, I can understand how this stored energy creates imbalances between the physical, emotional, mental, or spiritual components, resulting in diseases. As I studied each case of physical, emotional, or mental imbalance referred to me, I began to see the thread of specific energy patterns, and how they were transported through the various parts of the body. At the time, I was unaware of the name associated with the "vehicle of destruction" that traveled throughout the physical body, able to create debilitating illness and disease. One day, while working with a physician from UCLA Medical Center, I drew a picture of what I could see using my intuition. At last, I discovered the name: neurons, live cells.

The Science Part—The Neurons!

HOW IS THE BODY FED NEGATIVE OR POSITIVE ENERGY?

The answer is neurons. So what are neurons? They are cells that live within the human physiology. It's not necessary to understand science, medicine, biology, or physiology to grasp the concept you are about to read. I'll make it as simple as possible. Billions of neurons move throughout your body carrying information! The neurons have impulses, active energy components, memories, and power to create illness, excitement, stability, fear, excitement, and so much more!

Throughout the day and during sleep, neurons are being fed data, thoughts, and feelings. Awareness and action are crucial to maintaining high levels of powerful, passionate, confident data moving through your system.

The Good News

The neurons are waiting for your signals, thoughts, and photos to assure you that your hopes and dreams are within reach!

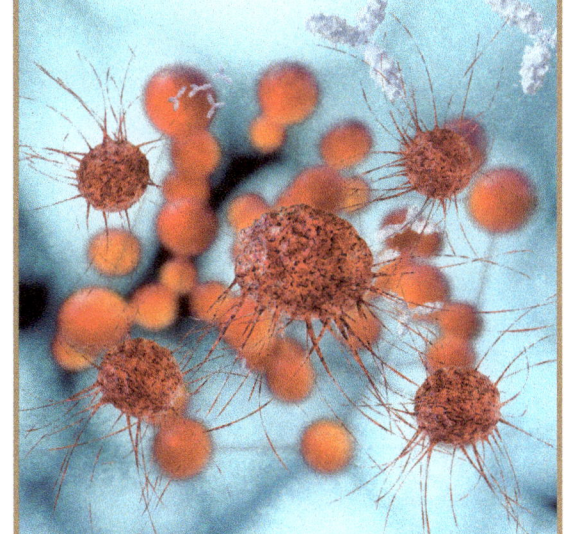

Each day you can recharge and reestablish the energy running through your body. The power running through your brain (thoughts), your heart (feelings), your stomach (knowing), and your eyes (memories) can be repatterned at any moment. The best news I can give you, based on years of success with individuals who "shifted" the energy running their bodies is that *your body wants to be healthy*! Neurons are traveling throughout the body *waiting for* signs and signals to bring energy, joy, and happiness through the channels.

The physical body does not want to live with pain, confusion, or frustration. If you begin the journey to realize what is happening within your physical body to be more in control of your health care, *bravo*! The next step after realization is to make an action plan. *Your* action plan; an action plan that can work for you within the frame of how you live today.

Create an ongoing dialogue with yourself today! Make those "snapshots" from your retina *count*! Bring in some joy, love, and creative stuff to make new, lasting memories. Listen to your stomach and your heart! It will begin the journey of repatterning the billions of neurons running through your body!

Epidemic with Emotional, Mental, and Physical Challenges

In my opinion, one of the most difficult challenges we face in the United States is the number of individuals who rely on medication for anxiety, depression, and high blood pressure. The numbers are startling.

Awareness, learning, and the desire to become more balanced are the first steps to making the necessary shifts in your mind and your physical body. Most important in your decision making—listen to your body.

- Depression medication is used by one out of every eight individuals
- Anxiety medication is used by one out of every six individuals
- High blood pressure medication is used by one out of every three individuals
- SSRI medication is used by one out of every four women forty to fifty years of age
- Sixty-plus million dollars is spent on prescription medication annually
- Sixty-plus million dollars is spent on holistic options, alternative medicine, and natural products

Nervous System Overload

What Causes the Human Spirit to Collapse, Sending the Physical Body into Despair?

- Malfunctioning neurons
- Unresolved grief and loss
- Anxiety, fear, and confusion
- Unresolved anger
- Excessive mental activity
- Emotional distress
- Feelings of being overwhelmed
- Loss of stability

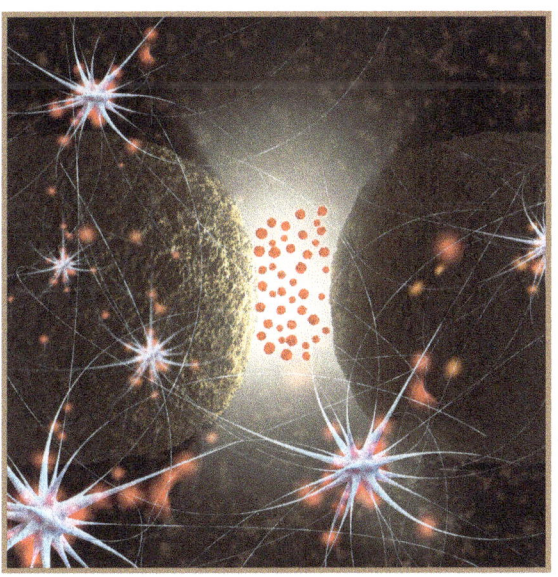

▲ Malfunctioning neuron

When a person experiences too much emotional distress; does not recover from grief or a loss; or experiences too much anxiety, fear, or ongoing confusion, the physical body will respond. If there is deep-seated unresolved anger or excessive, consistent mental activity without relief, the brain, emotional self, or the physical body will react. How will it respond, the physical body? Differently for everyone. Individuals are individuals. It is impossible to know how each person will or will not recover or return to what was "normal" after overwhelming life challenges. Seeking help when experiencing symptoms will result in many options. There is traditional (allopathic) medicine, naturopathic (holistic—natural) medicine, Chinese (acupuncture—herbal) medicine, Ayurvedic medicine, chiropractic, and energy medicine. All of these methodologies are teachings attached to a theory, on a belief and a promise that they have the answer to cure, balance, and heal the human mind, body, and spirit. Searching for the right therapy or medical attention in today's world of choices can be so overwhelming for some, and other can feel excited and passionate about the journey of learning.

How can some individuals heal while balancing their bodies and minds, while others with the same symptoms and diagnoses are staying behind? Based on the past twenty years of working with disease and debilitating emotional and mental imbalances, I say this with the utmost confidence: The energy to heal is different for each person. The human spirit, the will of oneself, and personal destiny, all play an essential role.

This energy, the personal spirit energy, distinguishes who wins the fight and who loses the battle. If you can push forward with consistent pattern changes, awareness being at the forefront of your thoughts, your neurons will undoubtedly take that data to the areas necessary for regeneration. If you cannot make the pattern changes, if you are too tired, if you sincerely believe that life cannot get any better, cannot be more satisfying, or that the physical body cannot become healthy again, it's going to be hard to change the energy of the neurons. This belief will be transported to the neurons circulating inside of you over and over daily.

In the year 2000, I went to an integrative medical convention on the Big Island in Hawaii. One of the guest speakers was a distinguished healer. His name is Papa Henry Auwae. During his speech to a standing-room-only crowd, he communicated to the doctors in attendance his belief about healing. He firmly believes that overcoming health challenges in today's world is 20 percent medicine and 80 percent spiritual (emotional/mental).

When I started my journey working as a medical intuitive, it was critical to me that I establish a solid factual foundation using my gift of intuition. I felt that offering my services in the area of medicine would undoubtedly weed out any inconsistent, false, or untrue information. Medical professionals want to be right. Medical staff members and scientists would undoubtedly let me know if my intuition were not up to par. This experience supplied me with the self-confidence to move forward with my gift. Working with various types of physicians, I discovered that they want to have the answers and feel that their years at medical school have earned them the right to have the answers. I cannot disagree with their knowledge, and yet, I can't help but wonder why millions of dollars is consistently being spent annually on alternative medicine and self-help modalities.

The area that concerned me during this period of my work was the differences in the opinion, data, and recommendations that are inconsistent when working in the various areas of medicine. There are so many options, so many views, and so many disagreements concerning what is healthy and what is not that it can be confusing for the patients. I believe this is one of the reasons that it's difficult for those seeking answers to be calm throughout the process of improving their health and wellness. There are many contradictions in the field of medicine, on a global scale. It can be very confusing to the patient and the patient's family.

With children, always remember, their nervous systems, little hearts, and brains are fragile. Whatever goes on inside the home, at school, and with their friends, the energy is magnified emotionally, mentally, and spiritually. If an adult struggles with emotions, mental pressure, and interactions in their lives, imagine what a child feels.

> **CAUTION:** The human spirit and the human physical body can only handle so much mental, emotional, and distressing life experiences before the nervous system, the heart, and the stomach begin to show signs of distress. The medical platform can see this disruption in patients with the following symptoms: anxiety, loss of appetite, exhaustion, obsession with fear, insomnia, forgetfulness, suicidal thoughts and lost connections to the world, depression, sleeping for days, and more. Often it can be a task for the physician to identify where to start, and medicines seem to be the fix for the millions of people suffering. I absolutely agree that prescription medication has a strong place in our society today. I also agree that we must provide options for forward movement to those who desire a drug free life after trauma or a life set back.

The Stoppers—Stay Down

There is a phrase I use in my work: Stay down! It refers to a life moment that is what I call a stopper—any moment in life that requires you to stop, pause, and process. No one knows the truth about the magnitude of what you are experiencing, even when you use words to describe it. Life moments can be attached to spiritual work, childhood traumas, or the evolution of your life experience. The stay-down moments are the moments when you feel as though you have had the carpet ripped out from underneath your feet and you have no way to get to a rail to hold you up.

I discuss these 'stopper-stay down' life moments in my workshops. They are the life moments that leave you crippled mentally, emotionally, or physically, unable to continue with life as it was. The moments in life that you make calculated decisions to continue the current ongoing battles in your life to keep going, or you realize that you've lost 60 percent of the energy you will need to continue with the struggles, and you put your sword down. The moment you decide to put your sword down, knowing that you do not have access to the physical and mental energy to continue onward, is the moment you are on your path to recovery. It is the moment of surrender. The moment you do not push yourself any further, realizing that you are not in a position to move forward with what it takes to win, complete a task or to remove yourself from the potential of a complete breakdown or collapse.

The "stoppers" should allow time to reevaluate the importance of priorities in our life. The stopper moments arrive in our lives during moments when we must evaluate the road we are traveling, discover and connect to what is essential in our lives, our desires within, and decide how we want to use our energy moving forward—a self-evaluation period.

Life Setbacks Require Internal Assessment Using Personal Wisdom, Intuition and Self Confidence

The following information is from my perspective, based on the needs of clients who have trusted my knowledge and support over the past two decades. Again, my job, using my gift of intuition, is to solve the mysteries behind tough-to-understand conditions wherein these hidden challenges may be attached to deeply rooted imbalances. I am not a medical professional. Instead, I use my keenly accurate intuition to uncover and understand the root cause of an illness or imbalance with particular attention to how it opens the mental, emotional, and spiritual doors allowing a platform for disease or imbalances to occur and grow.

One challenge I have seen an increase in repeatedly over the past ten years is the diagnosis of bipolar disorder. I believe the diagnosis of bipolar has increased by 50 percent in the past decade. Those diagnosed with the—label of bipolar often carry a rational thought filled with a heavy load of inadequacy. They can feel broken, embarrassed, and disconnected from normalcy. I have observed consistent findings among individuals I have had the honor of working with that have enabled me to solve the "reasons" that their emotional and mental bodies have broken down. Similar to a wire snapping, the neurons were no longer able to hold a connection, and the flow of the energy to and from the brain collapsed. The neurons simply deteriorated, reduced the electrical current, or broke off completely from the pathway. This phenomenon can occur when one experiences too many emotional life experiences or the rational mind has completely collapsed from overload.

The mental, psychological, emotional, or neurological energetic functioning snaps or, in some, becomes disconnected from what was previously normal. The weak area is no longer able to hold the necessary connection to keep the energy, the neurons, flowing. I've seen this too many times when the rational mind cannot control the emotional suffering for even one more minute. I believe that when the brain can no longer handle stress and the heart can no longer withstand emotional pain, the body malfunctions. When the body says, "No more," it actually will support the disconnect by breaking down. In some, it's the only way the individual will take a break.

Many clients I have seen have suffered severe life setbacks because of situations that occurred, either one big blow or repeated patterns. Sadly, it is rarely the one setback that breaks the physical body. Imagine a worn-out tire. During the wear-and-tear period, the driver knew that the tire had a certain amount of traction, or shelf life, under warranty. The body knows when it begins to malfunction, when the wear and tear begins to show internally or externally. When you stop washing your hair as often as normal or you stop answering your phone altogether or indeed with vigor, you know something is off, worn out, or worn down. The brain and vagus nerve send signals to the main areas of the physical body when the energy levels start to shift or shut down. If the signs and messages are not acknowledged, the physical body will malfunction. Like a tire, if you do not watch the tread and pay attention to your warranty, you can have a blowout, which would potentially end with much more damage. Life moments that stack up and cause malfunctions are just that: moments. The life lesson can be to become keenly aware of what you can handle, and what will completely break down your spirit and your physical body. The physical body will give signals when in distress. Pay attention!

KNOW YOUR PERSONAL TRUTH. FOLLOW THE PATH THAT FITS YOU

Allow me to shed some light on a scenario that would often be diagnosed as an anxiety disorder or a social disorder if the practitioner did not have experience working with or interacting profesionally with this type of individual. A diagnosis of asynchronous brain development is not common, so many physicians do not understand how to treat this type of individual. Again, another label. As a patient, this person can show behavioral signs that include intellect over excitability, profound creative development action plans, and passionate display of razor-sharp focus on current projects, often to the exclusion of anything and everything else around. I have had the opportunity to work with dozens of these individuals and feel quite honored. These are individuals who could be geniuses if tested, in many areas of music, art, mathematics, sciences, sports, and others. Because physicians in mainstream medicine do not often see these individuals, these individuals may take longer paths to wellness and wellbeing. They should not be diagnosed with the same measuring tools as the "normal" patient. Their spirit, their intuition, is much different than an average person's.

KNOW YOUR TRUTH, YOU ARE AN INDIVIDUAL.

I have found similar traits and energy in ultra-successful men and women in business, leadership roles, or athletic successes. They excel at the top of their games and, for that reason, they stand out because they have abilities that are above most others in their fields. The goals these individuals set are sometimes unimaginable for others, and yet, in these individuals, they are natural and matter-of-fact. Their brains, their intuition, their available energy cells, all work together to arrive at their desired results, almost always with little to no interaction with others. I have learned that their minds and their focused energy can work with little rest, and are merely traits of who they are as human beings and how they operate naturally. Often these individuals prefer to sleep four to five hours. They do not like to waste their life energy. There are physicians who believe this is unhealthy, and yet for these individuals, it's the "right" way to live and operate.

At times, these individuals are often not seen as "social" beings by mainstream society. They may attend the necessary outings, but they prefer to be at home, in the environment that feeds them and allows them to be relaxed. This type of individual is not looking for outside stimulation or acceptance. Most of them have a few trusted friends around them, and they are content. They choose not to fit in, not to be part of a crowd, and in actuality, to some, may not appear "normal," but in fact, they are viable human beings with their own sets of extraordinary traits.

ONE SIZE DOES NOT FIT ALL as "they" want us to believe, regarding emotional, mental, social, spiritual, or physical inner and outer appearances. Human beings are individuals; they are not supposed to fit into only three sizes of acceptable behavior or development.

Various media publications follow and focus on the spectacular individuals who go beyond the norm in their fields to emphasize how they are able to surpass the "ordinary" individuals in their areas of expertise. Why is that? Why is the media continually needing to dictate what is special, spectacular, unusual, genius, and extraordinary?

We each have access to an energy source attached to our spirit when we come into this lifetime. My biggest motivation is that you, the reader, discover how to open the connection to yours, to find the area of your strength and use the neurons to bring balance, excitement, and joy to your life. Feed your neurons the energy that you want in your life today! As often as you can, let go of thoughts of yesterday unless the memories have ideas of joy, happiness, passion, hope, and the feelings of being loved and loving others.

I address more on the issues relating to the malfunction of the nervous system, brain, and vagus nerve, further discussing my thoughts on anxiety disorders, Parkinson's, ADD-ADHD, and

Alzheimer's, including the prevalence of misunderstanding the signs and signals and use of medication, in my follow up book, *The Energy of Illness—The Energy Associated with Disease and Illnesses*. *The Energy of Illness* dives deeper into specific illnesses and diseases, using the neurons as the tool of destruction.

Be careful of the world's idea of what is and is not normal. What is normal? Whatever the "normal" standards are today, don't worry, tomorrow "they" will come up with a higher bar or a new theory. If we learn to be confident in our behavior, our wants and needs, without asking for acceptance, the energy circulating throughout the Invisible Anatomy (eye, brain, heart, and stomach) will remain calm and excited and can produce unlimited amounts of energy to live your hopes and dreams.

Without Conscious Effort, We Feed Our Body Poison

KNOWINGLY AND UNKNOWINGLY, OUR THOUGHTS, MEMORIES, AND FEELINGS ARE POWERFUL

Energy, and how we use it, see it, feel it, and apply it, can hold the key to many unanswered questions. To understand the effects of energy, you must first understand its essence. How you interact, react, respond, and engage in your life can be easily tracked. By watching the ongoing energy levels flow throughout your day, you will know how the energy of your life fluctuates and affects you. You are the best judge and understand how you feel and what your physical body's reactions are.

I believe that human energy is the essence of whom we are as living beings. Our thoughts (rational mental thinking) and emotions (feelings) interact within each of us, every moment of our lives. Even those committed to lives of solitude have this experience. Our spirits are born into this world and live within these physical bodies.

Our minds (thoughts) and feelings direct us from the moment we awake. Think about that. Please sit with that for a moment. When you wake up, how do you plan your day? What is the first step? It is a thought, feeling, memory, or action? And then we are on our ways. Our thoughts have an energetic effect on the people in our lives as well as an effect on ourselves. We use energy positively or negatively on a daily basis. It is that simple. We use it confidently to benefit our persons and those around us, or we can, unfortunately, use it negatively, often without knowing.

Many of us continue to repeat the same patterns, never entirely accessing those feelings we desire, obtaining the goals we set, committing to the diligence necessary to maintain healthy lives or to find the real success we seek. For many, thoughts, expectations, and intense energy can create self-sabotage. Without needing to analyze the reasons why these patterns and hit-or-miss moments can pile up, they do. They lead to disappointment, frustration, anger, and, eventually perhaps, illness. For some, it's tough to maintain positive momentum.

Energy Works for You or against You

Be mindful of your thoughts every day and be in a state of awareness throughout the day. It can help you with setting up a pattern of consciousness if you are working on becoming stronger in your passion for life, creating wellness in your physical self, or inviting more love into your life. If you have repeated thoughts, as most of us do, you may already be aware of how they can work with or against you.

Typically, we feel healthy, prosperous, and balanced, or we feel tired, overwhelmed, and stressed. For most of us, maintaining stability in our positive thoughts and feelings can sometimes be overwhelmed by the enormous stress and responsibilities carried throughout our lives.

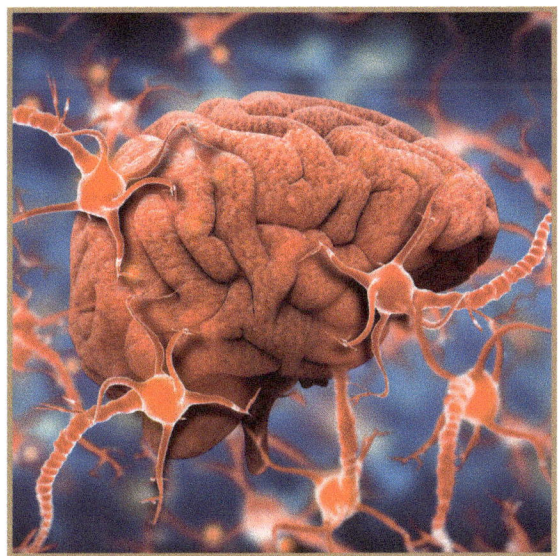

When we are overwhelmed with the now, and the future, both immediate and long-term, the energy opens the door to insomnia. Add in a cup of "past" anxiety, stress, or despair, and you've just mixed up a crippling cocktail for your spirit, physical body, and emotional self.

I have found that many of our conscious thoughts aren't always the culprits of our illness or disease. Many times, it's an unconscious thought attached to old feelings and memories. Within the consultation walls, I've heard the phrase over and over, "No, Jennifer, I've worked through that." Most of what these individuals thought they had worked through is still very much alive, under the layers of positive thinking! It can be very confusing.

In 2010, I learned something as significant to me as it was to the young woman who walked into my office. Isabella arrived at my office with her mother. She was not able to continue her last semester at Princeton University and needed answers to why she had not been able to have any relief with her health condition over the past year. She made the journey from the East Coast to see me in my office north of Los Angeles. Isabella was filled with hope to find a solution to her challenges, and return to Princeton University to accomplish her goal to graduate. Her debilitating condition (anorexia) had taken a toll on her. She was no longer able to stand up for long periods without support.

When she asked the following questions, I could not take my eyes off her beautiful face. I listened, and I too needed to know the answers to, "Why am I not healing? Why don't I feel better? What is wrong with me?" She continued, "If I have followed the guidelines outlined by the doctors, therapists, and reiki masters, why am I still sick?" I shared her pain at that moment. Her parents had spent thousands and thousands of dollars trying to help her find a cure, an answer to her illness. These were viable questions that deserved a complete and thorough explanation.

I set out to teach her why the healing, balancing, and forward movement was not a generic undertaking. I began our journey that day with my understanding of the memory tape, the "tape" that stores information from the day we are born. A key point for me at the beginning of our conversation was her mother's inability to allow Isabella to answer for herself. Understanding the sensitivity of the dynamic between these two women, I proceeded with care as I navigated the question of being "allowed" to sit with Isabella alone.

When I asked her mother if I could speak to her daughter alone, she was not comfortable. Her arms immediately crossed over her upper stomach and chest area. An automatic response of the crossed arms near the abdomen or chest is indicative of a person thrown into the emotion connected to fear! As with many entanglements of parents who have caused emotional and mental distress in the spirits of their children, she was nervous. Knowingly or unknowingly, some parents, unfortunately, are not capable or willing to allow their children to tell their version of what has gone on within the family units. Most parents believe they have done the best they can and that's the story.

Both were fearful and broken; a mother with a daughter who might not make it to her next birthday and a young woman knowing she was in serious trouble physically. I will always believe that after years and years of searching, trying every medical modality possible, [something] led [past tense] them to me, the last ditch effort to save her life.

During our time together, Isabella discovered that several events during her teenage years had caused her to doubt her self-confidence, her physical beauty, and her ability to communicate her feelings to those she loved and desperately needed love from, time and time again. The anorexia challenge set in at seventeen years of age. Her parents lived lives and interacted in a world driven by excellence, as they saw it. Accomplishments, visual beauty, and monetary success were a few of the "musts" in their home environment. Isabella's spirit had a kind personality and empathy for others, and could not keep up with the hard-driven expectations.

Instead of referring her to a physician on my list of qualified MDs, I recommended to her mother that she take Isabella to the nearest emergency room, close to their home in Los Angeles. I explained that I knew two vital organs were in great danger. Her kidneys and neurons to her heart were damaged and needed immediate attention. When she left my office that day, she took a bracelet "of awareness" for connecting to her heart, a journal for writing "new" stories of hope, and my prayer for the strength she needed to reconnect to her spirit energy. I shared with her and her mother a concise understanding of the months and years certain events had occurred that had broken down her spirit and the vagus nerve, and the exact day/moment her brain rejected the truth about food and weight. The day she created the disconnect between her brain's ability to know that eating was a necessity and not a danger.

I encouraged her to begin the process of "new food for the neurons" immediately and to continue throughout the day and days ahead. I wanted her to feed her precious body with "new" nutrients and pictures that would travel to the critical areas to help her begin the long process of healing. Meeting with Isabella changed my perception of debilitating illnesses and set me on a path to learn and discover more about the brain, vagus nerve, and neurons.

Support the Neurons in Your Body with Awareness

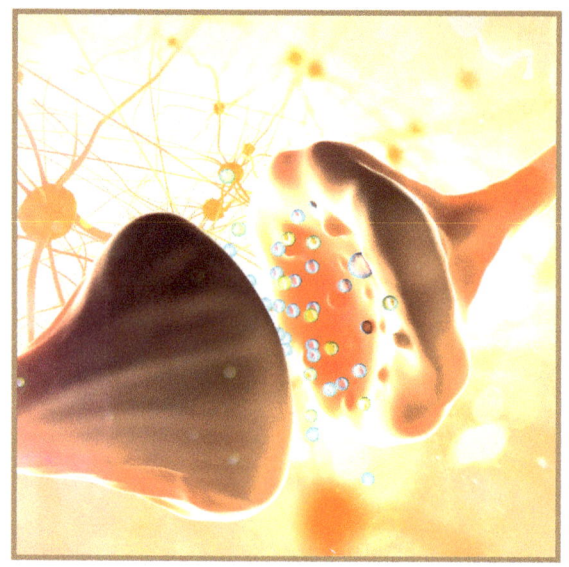

YOU HOLD THE KEY

The road to filling your physical body, mental mind, and emotional self and connecting to your spirit is a personal choice. It's a personal journey. Your body wants to live and live in balance, I can assure you of this. Only, however, if it's your desire. Many people do not feel the need to make changes in their lives, nor do they want to expand outside their comfort zones. Each person has the control to make life choices. However, if you are a person who would like to find a better way to interact in your life, you are reading this book for that reason.

If you need or want to make shifts to your stored memories, spend at least twenty minutes three times a day feeding your neurons new extractions from days gone by that were enjoyable and filled with hope and when you had some fun. Focus on those special moments that brought some joy, happiness, or feeling of confidence to you. Even in the darkest times, our spirits will bring in moments that send light to us.

If your goal is to connect to hope and excitement regarding a new option, opportunity, or relationship in your life, make sure that your body is on board and in tandem with you. Ensure that your thoughts, words, and feelings are positive and living in the moment. Live in the moment and confirm the reasons why you can accomplish your goal. Feed your system with the food you need to support this forward movement. Let's refer back to the tread on the tire. Be aware that things will wear out if you do not take care of them. Your body will send you signals, signs, and you will sense when you need to pay attention. When the time comes that various areas of your body are changing and shifting because of age or life experiences, take the time to check in and feel your way to the next stage of your life.

Remember that your spirit is with you on this journey. You are indeed not alone!

What I Know To Be True

The physical body wants to live a life filled with
PASSION, HAPPINESS, LOVE, and EXCITEMENT

A spirit lives inside the physical body
waiting to SUPPORT YOU

The human body is a miraculous vehicle
with EXTRAORDINARY POWER

INTUITION is real and a natural component
of the HUMAN SPIRIT and physical body

ENERGY OF LIFE SYSTEM™

- Wisdom Energy Center
- Intuitive Energy Center
- Communication Energy Center
- Love Energy Center
- Self-Confidence Energy Center
- Passion Energy Center
- Transmitting Energy Center
- Stability Energy Center
- Grounding Energy Center

BY J.K. DICKINSON

Revised in 2015

The Energy of Life System

CHAPTER 2

Let me touch lightly on the Energy of Life System. I created this system in 1998. Before embarking on the study of the neurons and how they affected the growth of disease in many of the prevalent health conditions today, I intuitively saw these energy transmitters landing in strategic areas of the body differently. I saw what I believed to be "circulatory clusters" of energy. This energy did not allow for smooth blood flow, cellular flow, and the necessary power to flow consistently from the brain to the lower legs. The electrical connection, the stream, stopped, planted itself, and caused destruction in the physical areas of the body. I found it fascinating and wanted to know more.

Although there is, in fact, a similar theory regarding these strategic locations, I felt that the data available for chakras was incomplete concerning what I had witnessed during consultations and research. It was in late 1998 that I created the Energy of Life System. The Energy of Life System has been successfully used by licensed medical professionals to follow and track patterns of energy that continue to create ongoing uneasiness in individuals.

I want to briefly review seven areas of the Energy of Life System with you. Although there are nine in total, let's begin this brief explanation with seven. You will find reasons for imbalance and a story of a client who experienced challenges in this area. We will tie it all together after you've reviewed the outline for the energy centers. The neurons run through these specific areas of the body. The vagus nerve is an enormous influence. When there is congestion, repetition with volatile emotions, a sense of being overwhelmed in mental chatter, inflammation with foods not easily digested, and so on, the breeding ground begins for illness, imbalances, and the inability to gain control over the physical challenges.

Stability Energy

CONNECTING TO CALM, ACCEPTANCE, AND FORWARD MOVEMENT

Life Experiences or Feelings That Can Create Imbalances

- Death of a child
- Divorce or separation from a loved one
- Introduction of a child or children without planning or mental, emotional, or financial preparation
- Denial of creating a family which includes children with someone who is unwilling to participate
- Child or children leaving home
- Sharing custody of a child or children after divorce
- Career loss without new opportunities or acknowledgment of contributions
- Mental and emotional abuse within the home environment
- Feeling low self-worth from your thoughts or mistreatment by others
- Multiple people parenting a child with opposing belief systems
- Loss of a loved one without support to move forward

- Addiction at an early age
- Aging without a feeling of purpose
- Unforeseen health challenges affecting you or a loved one
- Inability to fully recover from the loss or death of a loved one
- Undergoing abusive behavior by a spouse, child, or significant person in your environment, causing despair, depression, and damage of self
- The feeling of uselessness in your life, a life without purpose according to others

When an imbalance occurs in this area of the physical body, a person can feel ungrounded, confused, or disconnected from energy to move forward with confidence. If the imbalance proceeds, the physical body can begin to shut down energetically. There can be a feeling of exhaustion, a disconnection from activities that were once fulfilling, loss of smiles and laughter, and the overall sense of defeat.

If you or someone you love has experienced any of the life events above, please take care when moving forward with them. Assure yourself or your loved ones that the time to heal and rebalance is an individual journey. If the time necessary for processing a loss has not been completed, the imbalance in this area can be long lasting. Each person has an internal clock for processing life events that require a pause.

Client Story

STABILITY ENERGY 2002

Keko's Story

In late 2003, I received a phone call about a young man who was experiencing difficulty with many physical challenges. When Keko and I met, he had already undergone two major knee surgeries. His parents wondered if he would be able to continue participating in sports, as he was a very active boy. He also had insomnia.

My immediate priority was to help him rest. Keko had not had a peaceful sleep in over ten years. Insomnia affected his thinking and emotions. The spirit of the young man could not find a place of comfort, and so his physical body was unable to receive the nurturing, healing, and rejuvenation necessary to restore his system.

After several weeks of work, I realized that during the early days of his childhood he had become severely stressed over the dynamics within his home. He had a broken family with a highly dysfunctional father and a mother who had "ego-blaming" based energy. She believed that it was everyone else's fault that her marriage was over, instead of taking responsibility for herself. Sadly, she blamed her son's behavior for her husband's transgressions. She shifted her duty as a parent by not accepting accountability. His gentle energy took responsibility for his mother's downfall, graciously and without confusion or conflict. He desperately wanted his family to work.

Keko came into this world with a uniquely different type of energy system than his mother. He was empathetic to the point of storing every unhealthy event from his mother, father, and younger brother's lives within him. When I looked into his system, I could see that he was carrying everyone else's pain. He transferred the pain, agony, sadness, and anger from his parents into his energy system and body. The burden that this young man felt was incredible. There was no question in my mind that his inability to sleep came from all he worried about, stressed over, and accepted as his unbearable life challenges. I had never witnessed such a severe transference. His spirit was naturally integrity driven. That of his parents was not.

This young man's energy was that of a sixty-five-year-old, the approximate age of his parents. The first part of our work was gentle. With absolutely no blame or reference to his parents, I wanted to help him see what was honestly going on, so he could understand that the energy he carried was not his responsibility. He needed to learn that his parents' pain was not his own. From the age of three, this young man had tried to fix and mend his father's addictions and his mother's anger and blame. The guilt he felt was beyond any emotional word I can use. The pain he felt in his heart was apparent each time we spoke. It was hard for him to hold back his tears, while his beautiful smile still shone through during each session.

While we worked, Keko discovered that he had suffered two massive hernias. He underwent surgery to repair this immediate challenge. Again, it revolved around his lower extremities. Although he was doing his very best to overcome the difficulties stemming from the energy that was not his, his spirit was also involved in the drama. The sadness and the anger of the broken family dynamics continued to weigh heavily in his heart and mind. Keko was no longer able to handle the stress of his family, and his body showed it. His lower extremities were unable to carry the heavy weight of his family's issues, pain, and conflict. His body began to collapse slowly. His stability energy center was losing its strength; he was severely unbalanced.

Keko and I worked together to unwind his unhealthy connection with his family issues. Our desire was not to remove him from his family unit. Our goal was to move him into his energy, supporting his thoughts, dreams, and hopes. When he returned to playing the sport he loved, he once again felt pain. This time, it was his foot that had become severely damaged. His family searched and searched for the answers to the new condition. Nothing appeared on the CT or the MRI. This challenge lasted over a year.

When a person is stuck inside himself, without the resources or understanding to move forward with confidence and clarity, the physical body will carry a sense of helplessness—a feeling of internal despair without understanding the source. As time went by and Keko focused on his challenges with honesty, he began to wake up. As he made more associations between the emotions of the story and his body's interpretation and expression, he was able to surf again and walk firmly on the sand without pain. He began the road to integrating his new energy and how that would reunite him with the sport he loved, surfing!

During his self-discovery period, he realized, understood, and finally accepted that he was not responsible for the damage that his parents inflicted on themselves and the family unit. Keko realized that he did not need to carry the blame, nor did he need to fix what his parents had created. Through his self-discovery and long, long hours of preparing for what he knew his future could be, he started to self-evolve. He began to understand the core strength he held within himself. The shift was incredibly stable and formed a foundation for him to embark on the future he was designing within his structure—his ideas, dreams, and hopes!

Although the impact of his family's residue would never entirely leave him, Keko took on the task of genuinely healing and rebuilding the energy running through his system with his newfound love of his spirit and himself as a person. The work he undertook to balance his life, his physical body, and his core spirit was not about forgetting or forgiving. It was not about burying the past. He merely understood that his responsibility as a young man did not include caring for, counseling, or holding space for his parents in the dynamic that had been set up for him as a young boy. Keko was no longer the keeper of the family fallout. He was a young man ready only to accept responsibility for his actions, dreams, goals, and his idea of life and how it would work out for him.

Through his many walks on the beach, hiking, swimming, and daily centering of his energy on the earth, Keko learned to stand on his own two feet—the two solid feet that were able to carry him into the next phase of his adult life.

Today, Keko lives a comfortable life in a place he loves and calls home. His body responds to joy, happiness, and the rhythm he has set for his life. His journey has taken him to a place he was unaware was possible many years ago. He now helps others who struggle with challenges within mind, body, and spirit through his qigong practice and surf lessons.

During one of our consultations early in the relationship, Keko asked me about his future; specifically, if it was a possibility to find happiness and joy with himself. I saw that he would bring a young girl to his life, whom he would parent. In his current state of mind, he could not believe this. As time passed and eight years went by, he now has that precious little girl in his life along with a wife whom he adores. Awareness, intention, and committed action create a platform to live your dreams! It is possible.

> **CAUTION:** The parents of this young man did not accept their responsibility in inflicting their relationship challenges onto their son. Keko realized as he grew into his own power, focusing on his life, that he no longer needed or wanted their approval or acknowledgment of the past. When you finally allow yourself to let go, gain the strength from your own energy, know your life purpose and dreams have validity, LIFE WILL SHIFT!

Passion Energy Center

CONNECTION TO JOY, EXCITEMENT, CREATIVITY, FINANCIAL ABUNDANCE

Life Experiences or Feelings That Can Create Imbalances

- A man or woman losing employment, loss of responsibility, and income source
- A man realizing he is over his head in financial responsibility without resources to continue
- Losing a loved one to another person
- A woman losing herself mentally or emotionally after childbirth
- A woman waking up to domestic life with a feeling of numbness, despair, and loss of self
- A man or woman facing parenting alone, after a divorce or separation
- A man or woman realizing a secret infidelity during marriage to a life partner
- A man realizing the woman he married has disarmed his male energy through unrealistic expectation and aggressive criticism
- Discovering that your hopes and dreams are no longer in reach
- A man or woman facing a divorce without agreement
- Ongoing chaotic energy in the environment at home
- Realizing you are in a marriage or partnership without love, support, or affection

- A man or woman facing verbal, physical, or emotional abuse
- Losing credibility or promotion at work
- Losing intimacy with the person you love and desire

When an imbalance occurs in this area of the body, several malfunctions can occur. The Passion Energy brings life to inner and outward passion, the strength of creativity, financial abundance, laughter, and a sense of confidence in the sexual, intimate part of self. The energy of *passion* is what keeps us alive, excited, joyful, and filled with abundant amounts of energy!

If you or someone you love has experienced any of the life events above, please take care when moving forward. Assure yourself or your loved ones that the time to heal and rebalance is an individual journey. If the time necessary to reestablish the emotional/mental restlessness is interrupted, the endocrine system, the immune system, hormones, and the nervous system can begin the road to compromise. Passion is an essential part of the human spirit and human life.

Client Story

PASSION ENERGY 1995

Karina's Story

In 1990, I began work with a woman who had developed uterine cancer. She had learned through her physician earlier in the week that her cancer had spread to stage three. During our first meeting, I pulled out an empowerment tool I like to use: identification cards associated with the energy centers. This empowerment tool allows the client to use their personal intuition to identify their primary imbalance.

Without reading any of the writing on the cards, I asked her to select the Energy of Life card associated with the greatest imbalance she felt in her life at that moment. She immediately picked the passion energy card.

I was not surprised. It has been my experience that most people who have imbalances do have the answers within. The core challenge can be confusing because layers of other memories can mask the main event or moment that created the imbalance leading to disease. It is a tool that has designed a sense of empowerment for those seeking an understanding that can lead to their growth and healing. When the individual can unconsciously or consciously meet me at the door of the underlying reason for the illness, creating a plan can occur, and change can be rapid.

Karina knew deep inside why she was instrumental in creating the imbalance in the female organ area. She was tired, worn out, confused, and exhausted. When she married her husband right after high school, they both had high hopes for the future. During the first six years, her husband had moved the family around quite a bit because of work. Having two children at the age of twenty-five created an inner struggle in Karina. She had never experienced the pleasures and happiness associated with a stable childhood.

Now, as an adult, she was responsible for two children. She was not enjoying the pleasures of motherhood and felt disconnected from the outside world. Karina also felt discombobulated in the world inside her home and, most importantly, her inner self. She interpreted her experiences as feeling abandoned—alone without support from her husband. Karina never thought her husband had the same challenges since he had a job, colleagues, and entertainment associated with his career. Karina felt helpless most days, living inside a home she sensed was empty and without happiness. The struggle of raising two children by herself (even though her husband was home every night) brought too much pressure. She felt anxious and started smoking marijuana while taking antidepressants for relief.

Self-medication led to many new challenges. The life she once had no longer existed. She was not able to prepare meals for her children on time. She became so tired that she stopped cooking altogether. She started to purchase frozen food that she could pop into the microwave. She and her husband began fighting. He did not want his children eating fast food nor did he appreciate seeing her untidy on a regular basis. He became disenchanted with his wife's outlook and did not feel she was stable.

Karina's husband had a childhood entirely different from hers. His mother and father were not divorced, and they had set roles. His mother stayed home with the children, and his father had a career. His mother volunteered in the community and kept house. Karina had no idea what this lifestyle looked like or how to interact within it. She knew that she was not the type of wife or mother her husband required or expected. Although aware of her shortcomings going into the relationship years earlier, Karina had felt that in time, she could grow into the person he wanted her to be, along with the woman and mother she knew she could be. She did not wish the confusion or unsettling issues associated with her childhood to happen inside her own home. However, this was not possible. Karina did not have the life skills to fit into this environment as quickly as needed.

Shortly after the marriage ceremony, the children arrived, which they both had agreed on. They were both excited to start a family. Danny's idea of family and Karina's view, they later learned, were as different as the colors of black and white.

Before coming up for any air, two children later, Karina felt continually displaced, due to several relocations from job promotions for her husband. Karina did not have the courage she needed to settle into the natural family lifestyle her husband had created. After this honest evaluation of her life, Karina started to neglect her hair and her clothing choices. She was no longer interested in taking care of herself as she had years before. It was no longer a priority. She described it this way: "It's everything for me to remember to get three meals a day for my children. I don't have time to think about myself anymore."

Although on numerous occasions she tried to speak to her husband about the severity of her challenges, he was not able to hear her. He was too busy creating income to sustain the family. Her husband did not grow up in a family in which the mother discussed her feelings. Instead, his mom was the foundation that allowed his father to maintain his status in society. However, Karina was not Danny's mother. As he became more disappointed in her, his days at the office grew longer. Weekends started to blend into his weekly workdays. He disconnected from his family. During their eighth year of marriage, her husband began an affair with a woman from the office. He felt he needed to create some balance within his own life to maintain the family responsibilities. The energy inside his home was too much for him to handle. He wanted to run away. When Karina found out about the other woman, she was devastated. The illusion of her marriage and her life as a mother came crashing down.

Within six months, she was in her physician's office complaining of pain and abnormal bleeding. The CT showed a uterus covered in tumors. His advice was surgery and chemotherapy. When she delivered the news during our appointment, I sat with her and gazed into her eyes. I wanted to know which road she would take at this part of her journey. Would she let go from exhaustion and fear of the future, or fight for herself and her children? When illness sets into the physical body, it can have a profound effect on the emotional self. Depending on the immune system, nervous system, spirit energy, and conscious and unconscious energy, there is a fine line between giving up and having the strength to keep going. The strength of illness can hold a great deal of memory association.

We held hands as she talked about what life meant to her and what it might mean to her children if she were no longer there to love and support them. With this news, her children became the focus of her life. She needed to come up with answers to who would care for her children and how, if she were not available. Suddenly, almost shockingly, Karina found her voice. Her children became the joy and love of her life. This energy exchange, her openness to reconnect to her children, and their pure love and devotion to her happiness was the ticket she needed to turn her life around. She had always felt this connection was possible but had felt disengaged from herself. Now, from that point forward, they were a team. They were in this fight together.

Healing is a process. The teacher, doctor, or advisor can only take you so far. You will have the energy inside to push on, or you will not. Depending on the depth of your illness, there are times the immune system might not be able to make it back to normal. Listen to your body! Listen to the signs and signals it gives you! Karina's decision to push through required her to take a hard look at her life. She needed to find out what she had within herself as a young woman, regardless of her childhood. She needed to reconnect to her inner strength and realize what her priorities were and what she was able to do to create a place for healing within the walls of her current environment. With the help of her two sons, who loved her deeply and needed her companionship and guidance immensely, Karina decided to undergo surgery. With or without the support of her husband, she had two beautiful smiling faces greeting her each day with love and compassion, and a phrase she often heard: "Mommy, you can do it." She started to believe it within herself. Grabbing onto that energy, she stuffed it into her heart area and held on with dear life.

Karina had to embrace and acknowledge the experiences that were so painful for her in childhood: her mother's leaving her abusive father and living without the means to support a family as a single mother, and the hardship that endured because of those decisions. For once, the memories were a blessing. She realized that watching her mother survive the hardship year after year had given her the foundation to rebuild her own life with her two children. An experience she never felt could have provided her with anything but the pain was now the cornerstone that would help her move forward with her own life. Checking in with her months after the surgery, I was happy to hear that the changes she had made in her life allowed her more stability than she had ever thought possible.

Karina realized that even though she had experienced painful moments in her life and her childhood, her future was not to be determined by her past. Instead, it shifted and became stronger by the energy she put into her daily life. Today, Karina lives with a great sense of hope and self-confidence. She lives with an encouragement that life can continue to be an incredible journey, regardless of her childhood or the memories that followed. Karina understands today that painful and unhappy moments in childhood can, and did, come to her aid. She also realized that the fear, sadness, and despair, when viewed from a very different perspective than the one she had formed as a child, allowed her to persevere with the knowledge that there is light at the end of the tunnel, regardless of how dark the tunnel is or how long the pain has existed.

This energy center is linked to fun, excitement, joyfulness, close intimacy in relationships, building creative ideas, and maintaining positive relationships with those around you. Passion energy is also attached to personal relationships within a family unit. Sharing oneself with others openly, without restriction and reservation, is a benefit of a balanced passion energy center. Feeling warm, cozy, intimate, and honest with yourself and others, especially the family surrounding you, is a sign of a balanced, thriving passion energy center.

 ## Self-confidence Energy Center

CONNECTION TO CALM, ACCEPTANCE, AND FORWARD MOVEMENT

Life Experiences or Feelings That Can Create Imbalances

- Being told you are not capable of achieving things that are important to you
- Hearing that you are slow, stupid, and incapable during childhood
- Having difficulty during elementary school years
- Losing a promotion that you felt was earned

- Being in an environment that does not allow you to use your intellect
- Living in an environment that does not let you make decisions
- Being involved in social situations that create a sense of emotional abuse
- Being included in a family unit that does not support your dreams and thoughts
- Losing the respect of a loved one, one of your children, or a close friend
- Inability to execute plans or meaningful goals
- Hearing that you are not good enough time and time again
- Feeling that your body image is not perfect
- Being told that you could do better, be better, and think more clearly
- Being passed over for promotions time and time again

A balanced self-confidence energy center allows a person to feel confident in the social, personal, and professional arenas. A sense of accomplishment, association, and overall comfort with interactions, internally and externally, is a benefit of this energy center's balance. This helps create a person who is not selfish or self-absorbed, who can give without expectation, and who can give openly without needing in return. You are a person capable of sincere compassion toward others. You make decisions and take risks without hesitation or concern.

Client Story

SELF CONFIDENCE ENERGY 2010

Meredith's Story

Meredith came into my office complaining of stomach problems. She was often unable to hold down food. Her motivation to eat was gone, she no longer enjoyed eating. Every time she finished a meal, she felt anxiety. She bloated and suffered indigestion. She was experiencing a slowdown of her digestion, and a gripping sensation in her gut area made her feel uneasy and out of control. She also often felt severe pain lodging in her left upper stomach cavity, just under the rib cage.

Meredith was also exhausted, had panic issues, and was not able to get through her day without crying. We sat for a few hours discussing her uncomfortable symptoms and the issues that were upsetting her. As an experienced medical intuitive, it is essential at times to allow the person to communicate what they feel the real problems are, even though it may or may not be the actual reality. Truth is a very tricky word, especially when you tie in emotion, mental uneasiness, and the person's need to believe the facts as they tell them. Unfortunately, we can all create stories to help with how we feel. Although the stories are not always in direct conflict with the real truth, they can at times have added experiences or feelings that can complicate the truth.

As I sat and listened, I watched her hold her stomach with both hands. She was distraught with her husband, her daughter, her ex-husband, and the friends she had recently made. She always referred back to "better times" when her life had more fulfillment, more happiness, and joy. With vast experiences and feelings of pleasure and contentment in the past, she could not move forward. Those were the days she had not fretted or stressed over trivial issues.

We discussed her blessed life from twenty years earlier, and she became very excited. She did not hold her stomach during those conversations, and I noted how she talked with zest and high energy. I realized she felt that her life only indeed existed in the past and was not alive for her in the present. She was unable to embrace anything going on today. Something had happened that stopped her in her tracks, and she could not step into life as it unfolded from that significant day, that moment.

Several years back, Meredith had high hopes for a new and second marriage. Although she knew she had been feeling her intuition telling her that it was not the best move, particularly for her two children, she moved forward with the wedding. Meredith wanted to believe it was going to all work out for the best, for each of them. After being alone for eight years after her husband's death, she felt she had the courage necessary to embark on another marriage. Her first marriage had brought great joy and happiness to her and her family. At the time, she was not responsible for having a career or financial stability. She was a mother and a loving wife. The pride she felt as she moved through her life and moments with her family sustained her very soul. She knew how to support her children in a loving and carefree manner while gently supporting, caring, and loving her husband. Her world revolved around loving and supporting her family, those she loved.

After her husband's untimely death, she felt a need within herself to open the door to the possibility of uniting with another man. She did not want her children to live without a father figure nor did she want to live without a life partner. The fifteen years she had spent with her first husband only held beautiful memories for her and her children.

After being introduced to a man she felt would be a wonderful husband as well as a great father, she felt hopeful, passionate. Her responsibility as a mother was something she did not take lightly. Although, over the years, she had opportunities to meet with possible partners, she had not felt ready until the day she met Sam. There was something about Sam that created a sense of excitement that she had not felt since the passing of her husband. Sam, too, had been married once, years ago. There were no children from the marriage. After a few dinners and several evenings of discussions, the couple decided to embark on a life together as a family. Her wedding was undergoing the planning stage. There was hope for her and her children. If the next twenty years were like the past, this family was entering a new time of happiness, prosperity, and love.

When the children met with Sam, initially and on several later occasions, both the daughter and son had feelings of uneasiness. They voiced their opinions to their mother. They both felt that Sam was not his true self. Although they could not describe it in words, the daughter did say, "Mom, he's hiding something." Meredith listened with great care while she tried to uncover the real reason for the uncomfortable feelings they both had experienced. During her time with Sam, she did not see him as anything but generous, kind, and thoughtful.

Usually, she would have listened to her children. However, under these circumstances, the passing of her husband and being alone for several years without his warm comfort and shoulder to lean on, she felt that she needed outside support to help with a decision. She saw an advisor who told her that the children were suffering from separation anxiety and might not be open to allowing another man outside of their father to enter their lives. The advisor encouraged her to accept the marriage proposal for herself and her children.

During the weeks that led up to the wedding, light signals popped up that made Meredith uncomfortable. During an evening out in town, Sam drank more than usual. He became irritable with the waiter and raised his voice while discussing issues surrounding the children. Sam quickly caught himself and apologized. His excuse was the tension he felt surrounding the marriage, relocation of his belongings, and the expense of the wedding. He finally started to open up and voice his opinions about the future and his fears. While Meredith listened, she felt uneasy and agitated. She doubted her decision to marry Sam. She felt as though she needed to pull back and reevaluate the relationship.

The following day she called her advisor. He told Meredith that he felt her fears were unfounded, that she was getting cold feet. She was assured that within a few months' time the family would blend nicely together, embarking on a new adventure.

Several days prior to the wedding day, family started to arrive. Meredith heard the opinions of family members as they began to meet and interact with Sam. Some liked and accepted him, but Meredith's mother, in particular, did not like him at all and sensed that he was hiding his true personality. Meredith chose to move forward, a decision she accepted full responsibility for. She heard opinions and made the choice to move forward regardless of the signs and signals that were being shown to her. Choosing to ignore her inner feelings, her personal knowing, changed the course of her life—the life she knew and felt comfort within.

The day of the wedding, Sam was drunk. Meredith was horrified and surprised. The honeymoon was also not as she expected. She felt empty and isolated instead of loving and intimate. Sam's intoxication did not allow for any intimacy. What had she done? Although all the feelings surrounding the initial connection and attraction to Sam were still there, signs of difference and dysfunction started to arise.

A few months into the marriage, the children started to speak less, retreating into their separate rooms. The relationship with their new step-dad did not unfold the way they had hoped or expected. Slowly, the family unit started to shut down. Life, as these three individuals knew it, was over. Sam also started to shut down. Although he wanted to be the father and husband Meredith and her children longed for, it was not possible.

Sam did not like noise. His version of what noise meant was not apparent initially. As time went by, the children's voices and needs became "noise." Although this was not communicated at first, it was real. Sam did not want to face the reality of how he had lived his life the past ten years, or twenty years for that matter. In his mid-forties, there were many challenges he faced by uniting with this or any family unit. Many of the reasons that his first marriage had ended in divorce were slowly becoming alive again. Meredith did her very best to keep the lid on the volatile situation. But, soon after the marriage, the yelling began, as did the excessive drinking and the abusive language.

Meredith's self confidence, self esteem, passion, and wisdom, collapsed the day he entered the room and witnessed his daughter being shoved against the wall. Sam had a complete meltdown and took his anger out on Megan. This milestone in the marriage destroyed the balance of Meredith and her two children. The incident was so unfamiliar to Meredith and her children that the shock set up an emotional explosion that was not easily cleaned up. During the three years following this incident, the family was disconnected. Meredith was so distraught that she could not find a way to reconnect her family again.

Three years later, Meredith sat in front of me repeating the same story as she tried to remain alive in the past, unable to move forward with any sense of self-assurance. She spent two years trying to find herself again. Although Meredith had the self-confidence to make sound decisions from an early age, the relationship and interaction with Sam, and her unwillingness to listen to her gut feeling that things were never right, disconnected her from her self-assurance and self-confidence. Her body was expressing its pain and distress in her center of power, right in her stomach. Each time she looked at her children, the once very happy and contented loves of her life, now sitting lifeless and without joyful expressions, the pain got worse. It was even worse when she started to think of the past, remembering the beautiful times that were no more.

After the divorce, Meredith spent a few years in self-discovery. Her children were able to feel the essence of their mother again, although not the same as their childhood memories. Returning to the close family unit they all needed, the foundation was now being rebuilt on a platform very different from what they had experienced over the past several years. Meredith's children were older now, and they were able to see the painful outcome of decisions that their mother had made. Meredith was at the helm once again, listening and engaging with her children as she had so sadly missed, helped her regain her self-assurance. Rebuilding a strong family unit, with many life lessons to express to each other, was undertaken within this strong family bond.

To support Meredith in reclaiming her self-assurance, I developed a series of exercises to help her to change her patterns, step by step, day by day. I asked her to keep a journal of her thoughts and her feelings. I also suggested that three times a day she read her empowerment phrases out loud, into a mirror, making eye-to-eye contact. Also, I encouraged her to alter her daily life plans, little by little, with a firm intention to reclaim her self-assurance and desire, with the promise that through her commitment she would build the energy to support, feel, and connect to her new life goals. This included listening to her very powerful intuition.

Meredith suggested that some of her daily changes would consist of a slow walk in nature, conversations that were mostly upbeat, and the deep conviction that negative self-talk or "story energy" would stop as soon as she identified it as such. Walking clears the mind and gives the physical body a moment to oxygenate. Upbeat conversations provide an active connection to positive interaction. Her self-awareness of the unhealthy chatter—and then repeating the mantra of why that noise is no longer necessary—is strong medicine.

Love Energy Center

CONNECTION TO LIVING WITH EASE, ACCEPTING AND GIVING LOVE WITH A FEELING OF JOY

One of the most potent and destructive energy centers is in the heart area, love energy. The confusion around the word love, the feeling of love, and the giving and acceptance of love are more complicated than we can understand. During periods of life when there is grieving or feeling a loss, the pain in the heart can be debilitating. The word love and the feeling of love hold an incredible amount of power. The word love can feel like a double-edged sword for people who have been intentionally hurt by manipulation.

Life Experiences or Feelings That Can Create Imbalances

- Death of a loved one
- Abusive behavior from a loved one
- Abusive behavior to a loved one
- The feeling of love's being taken away without notice
- Destructive and abusive communication from someone you love
- Mother or father unable to enjoy their child/children
- Inability to participate in the life you want due to self-confidence challenges
- Failure to grow in a relationship
- Failure to be seen as the person you are, not were
- Death of a person without closure

- Loss of a person with guilt and unfinished words
- Death or loss of a person with regret
- Misunderstanding of the feelings of a person you've loved
- Being told you are not good enough
- Being told you no longer have a place in someone's life
- Sharing custody of children after separation or divorce
- Being replaced by someone in a loving relationship
- Being cheated on by someone you love
- Being discarded by someone you love
- Not being able to express your love openly
- Inability to know what healthy love feels like

Love energy, the heart of the human body, responds to two very distinct feelings in association with the word love. There are the pain and uneasiness from being removed from a loving situation without approval or notice, and the love, affection, and support given when exchanged with others. Loss and manipulation associated with the word love can be debilitating, crippling, life-changing.

The Stopper – Stay Down

Remember the phrase "stay down" earlier in the book? The heart has "stoppers" that are worthy of staying down for a moment or two. The heart can withstand pain, yes, and with time, can heal. A key point to remember, intentional motives to hurt someone, calculated moves to cause pain and suffering, or strategic manipulation to steal from someone who's opened his heart, can be the most crippling of all feelings a human can undergo. The result for the unknowing participant can cause challenges in the stomach hormones, brain fog, and so much more. The painful energy is running through the body with alarms going off.

When we are vulnerable by opening our hearts, and find we have been carefully, unknowingly, manipulated, the healing journey must begin with self-love. Love yourself enough to know it's not your burden to carry—a person who calculates manipulation, causing unnecessary pain to others, will have Karma knocking on his door. Karma has a way of helping those who need reminders. Giving your precious energy to someone who was and is not worthy takes from your spiritual essence, takes from your life energy and life experience and excludes you from the genuine interaction of life with another. Don't allow this energy to cripple you or steal any more from you! Do not feel it's your responsibility to teach the manipulator a lesson. Move on, get going, and grab some happiness!

Client Story

LOVE ENERGY 2008

Cleve and Cynthia's Story

Two Different Meanings of Love
The Athlete and the Model
Cleve was an incredibly successful athlete. In his early thirties, he was set and solid in his career. His father had taught him to get his life in order before he became engaged in a long-term relationship and

Chapter 2

Cleve waited until he felt ready and available. Cynthia was a young, forceful woman who came from flighty, unstable, divorced parents.

Cleve was the son of very stern, responsible, and thoroughly committed parents. Cynthia had difficulty in school and had a great need to prove herself in the world. Despite their essential differences, Cleve loved Cynthia, and Cynthia felt tremendous possibilities with Cleve.

A year into the relationship, Cleve asked Cynthia to marry him. Cynthia did not know that Cleve was so interested in her. She felt both disbelief and excitement. How could Cynthia, a girl from a broken home who was always on the move with her mother and who lacked a real sense of place attract such a successful and decent man? Was it all the confidence she exuded from winning so many beauty contests when she grew up? Was it her charm? She questioned her relationship from the beginning. Why me?

The moment Cleve placed the exquisite three-carat diamond ring on her finger, he was secure with his decision and knew that he could provide for his wife and future family. He was confident that his career would take him successfully into his thirty-seventh year, and he knew he would make a wonderful life for Cynthia.

Cleve sat down with Cynthia to go over the conditions of the marriage. His father had suggested he talk to Cynthia about his role as father, husband, and provider. He modeled his words after the severe talk he'd had with his dad some years earlier. He needed her to understand that his work took him away several months every year and that for the next seven years, he would be traveling during the season. She didn't seem to mind. Cleve and Cynthia planned to hire the assistance Cynthia needed to maintain their home. She was very hopeful and felt great pride in Cleve's status as an elite athlete.

Cleve was content. He had crossed all the t's and dotted all the i's, he thought.

Still intoxicated by her remarkable, custom, designer wedding dress and beautiful, big ring, Cynthia was floating on cloud nine. Cleve spared no expense to give his new bride all that she wanted. He had a great time seeing her excitement as she prepared for the big day. The wedding was everything Cynthia could have dreamed.

When it was time to purchase a home, Cynthia started to change her attitude toward Cleve's wealth. In the beginning, she was nervous about spending too much, did not want to overindulge, and did not want to overreach. Cleve was very impressed with his new wife's approach to finances and her lack of need to be with the in-crowd. However, because of his status, she was spun into the celebrity life very quickly. This life did not affect Cleve. He had signed his first contract at the age of twenty-one. He was well into his ninth year of significant, well-earned success. He was not interested in drugs or excessive alcohol. He enjoyed his career and wanted to settle down to the life of a married man, hoping to have a few children along the way.

The house the couple eventually purchased was well over the two-million-dollar price range Cleve had set. It caused a bit of discomfort since Cleve was a planner, but he wanted to make his new wife happy. Cynthia was so excited about the future with this incredibly successful man. She was basking in the glory of her newfound upscale life. When she started to choose the furnishings for her beautiful home, she learned from a new girlfriend that a decorator was the right way to decorate the house professionally. People of this stature do not do-it-yourself home decorating. The decorator she hired, without consulting Cleve, cost nearly $150,000. Cynthia rationalized the expense by stating that she did not have enough experience to purchase furniture for such a huge home. Cleve was not happy. He wanted his wife to participate with him in buying and selecting the furniture. This episode would be the first of many unsettling conversations between the two.

They enjoyed the first three months of marriage without much conflict. Shortly after the honeymoon stage, Cleve started to notice that Cynthia became somewhat addicted to buying new things; something Cleve was not comfortable or in agreement with. Without discussing with Cleve, Cynthia began spending large sums of money. Cleve sat down with Cynthia to address his concern. His words created an uncomfortable feeling within Cynthia. She started crying and demanded to know if Cleve loved her.

Cleve did not know how to answer that question, nor did he understand why that question had come up surrounding her overspending. He calmed her down with his gentle words. He explained to Cynthia that overspending was not a good pattern. He wanted her to fill her days with other activities. He was concerned that she'd become like his friends' wives: a massive consumer without any other real interests. He cautioned her to maintain the sense of herself she had before their marriage. Cynthia tried to listen but didn't hear the depth of what Cleve was saying. He wanted to warn her to avoid the traps of the entitled life, but she was unable to realize that. She withdrew from hearing the words that could have saved her marriage.

It was time for Cleve to set out on his traveling schedule. He would be on the road, on and off, for several weeks, and eventually, several months. Cynthia did not feel pressure as she had new friends who occupied her time. Cleve would call home to find that Cynthia was rarely ever there. She was out with her friends, dancing and drinking at all the hot spots. He found out through his friends' wives just how often Cynthia was out on the town. He did not like hearing the news. He called Cynthia often asking her to find other activities. She could not understand Cleve. Her credit cards were maxed out, after charging thousands of dollars on couture clothing, upscale jewelry, and handbags.

What had happened to his wife? Who was she becoming? Alternatively, who had she been all along? When Cleve arrived back home from a long working trip, Cynthia was not home to greet him. She was with her friends. Cynthia did not feel that it was necessary to reunite with Cleve the moment he walked through the door. When she finally came home later that night, Cleve asked her why she did not feel that it was essential to be there when he returned. Her response told Cleve that she was not interested in his return dates. He was at the end of his rope. He did not yell or use abusive language. He went to bed. Cynthia did not try to comfort him.

The following morning Cleve sat down with Cynthia to talk about their relationship. She didn't know what to say. She did not feel that her behavior was an issue. Cynthia's overspending, staying out all night, not answering phone calls, and not greeting him after a time away were enormous problems for Cleve.

Cynthia's response was simple, "What do you want from me?" He did not know how to reply. He had not expected this response. He had not realized just how dangerous the discord between his reality and her reality was. At this very moment, Cleve realized that Cynthia did not understand how a male and female related to each other in marriage, or that a committed, loving relationship operates on shared values. At that moment, his intuition kicked in. He knew there was trouble ahead. He wanted to help Cynthia understand how beautiful their life together could be, and he knew deep inside it might not be a journey that they would take together. He made plans to take her out of town for the few days he had before his next trip.

They had a wonderful time outwardly, but inwardly Cleve felt that Cynthia was different. When they were together intimately, it was not the slow, long, passionate process it had been in the beginning, during the planning of the marriage or even in the first few months after that. Cynthia wanted him to hurry up and get it over. It was not a pleasurable experience for her. Cleve felt that it was work for her. He wanted to experience that beautiful woman he had married, that tender touch, that sweet kiss, and the excitement that had been electric between them in the beginning. Unfortunately, it was not to be.

After returning home, Cynthia jumped back into the relationships with her party friends. She was content. Cleve was lonely and depressed and felt unloved. Cleve left again on a traveling work trip. He remained out of the home for several months—a friend of his heard from his wife that Cynthia was having an affair with a celebrity she had met at a famous club in Los Angeles. Cleve could not believe the news. He did not want to consider the possibility. He did not want to think about it or react to it. He needed to focus on his work. The calls to Cynthia were less and less frequent until finally, he just stopped calling. He hoped deep down that she would start to initiate the phone calls as she had in the beginning. Also, he remembered the time when she had confessed her love, the affection he had felt when they were intimate, and the overall feeling of hope and happiness they had spoken of when they planned their marriage.

When Cleve returned home, his wife was again not there to greet him. He sat in the living room with patience until she walked through the door. Drunk and unpredictable, she sat down to discuss a topic to which Cleve needed answers. Was she having an affair? Was she still in love with him? Cynthia had just enough alcohol to blurt it out: "Yes, I was sleeping with another man; yes, I am still in love with you; and yes, I still want to remain in this marriage." Cleve was shocked. He sat in fear, anger, and complete confusion. He did not understand what his wife was saying. He heard the words but did not understand the meaning behind them. He left the room silently. He slept in a spare bedroom that night.

When he awoke, he lay in bed trying to gather his thoughts. He needed to discuss the marriage, the relationship, and Cynthia's ideas of what was happening between the two of them. Cynthia was very clear when she started to speak. She told Cleve that she did love him, but that although grateful that he wanted her as his wife, Cynthia was not and would never be, satisfied with one man. She felt that her looks were a huge part of her identity. Her desire in life was to use her looks to obtain as much happiness as possible. Her idea of happiness was the affection and attention she received from men. It was the desire and affection from sex. Happiness also included the presents and financial options available to a beautiful woman.

He sat in disbelief. Cynthia tore out his heart with the words that had absolutely no feeling or emotion attached to them. Her words were very matter of fact. What he learned at that very moment was tearing his heart and self-esteem apart. He sat in his chair crying. Cynthia did not express much emotion. She asked if Cleve wanted a divorce. He sat staring at her for what seemed to be an eternity. "Yes," he said softly, "I would like a divorce."

Her next question was without any emotion: "Can you tell me how much money I can take with me?"

"Enough to live on," he said.

When Cleve came to my office wanting help to resolve the chest pain, stomach challenges, and heartache, I did not want to believe his story. I needed to hear it from Cynthia. The devastation was unreal for him. She agreed to a lengthy recap of her time with Cleve. My thoughts and feelings were with both Cleve and Cynthia. The story was such a tragedy. By phone, she confirmed the entire story. Even though I knew it was right, I did not want it to be true. Not for Cynthia, not for Cleve. Cynthia will remain, throughout her lifetime, searching for love, real love. She brought Cleve into her life to show herself, and experience with him, the true meaning of love. Unfortunately, she did not know what it looked or should feel like, long- or short-term.

Cleve had the affection, the stability, and the capability of creating comfort effortlessly. Because of Cynthia's loneliness and adverse childhood experiences, she was not able to embrace friendship, security, and support. She was still in the mode of going from relationship to relationship, trying to find the stability and love she had craved as a child. Although she had created what her heart truly desired, and could have stayed in a stable, loving lifestyle forever with the man who loved her dearly, she did not recognize and was not able to enjoy the energy of a calm and loving family. She had never experienced it and so could not appreciate it. Although painful, she could not abandon her past childhood experiences—she was only here for a moment before she moved to the next opportunity.

Because of Cleve's upbringing, he believed that everyone wanted the life he and his family had enjoyed growing up. He thought a woman would be proud and fulfilled to have such a man with his respectful, stable, and loving (from his perception) lifestyle. What he did not understand, and was not aware of, is that not everyone feels the love in the same way. Cleve learned through shock and devastation that love means something entirely different for each person who speaks the word.

A year later, I caught up with Cleve. He was still single, traveling with his work, and slowly recovering from the feelings of failure in his personal life. Cynthia, sadly, spent her settlement. She did not buy a home, did not invest, and no longer had her "friends" in Los Angeles.

Communication Energy Center

CONNECTION TO SPEAKING FREELY WITH CLARITY AND CONFIDENCE

Life Experiences or Feelings That Can Create Imbalances

- Fear of rejection, again
- Rejection during critical childhood moments
- Rejection during social situations in childhood
- Feelings of inadequacy compared to others around you during childhood
- Being left out of social circles during elementary, secondary, and high school years
- Fear of being judged again after moments of embarrassment
- Fear of not being accepted into the social environment you desire
- Fear of speaking from a place less intelligent than others
- Fear of reprimand or angered response
- Fear of losing your position in life, career, or relationship
- Fear of being misunderstood
- Fear of not being loved
- Fear of not having the "right" or applicable words
- The uncertainty of words flowing correctly with the intention
- The risk of speaking your truth and being attacked

- Inability to talk about experiences that have caused pain

The inability to communicate one's truth in life, relationships, at work, with friends, and most importantly, with oneself, creates enormous pressure on the heart and lung area. When words need to be spoken, and the energy is stifled, the energy that should be moving outward will be stuffed back and travel downward, and can be a breeding ground for imbalances and illness. Our words, thoughts, and communications are what allows us to live our lives with expression. The neuron track that becomes congested in the throat area will stop the flow of necessary energy moving to and from the brain into the lower part of the vagus nerve.

Client Story

COMMUNICATION ENERGY 1996

James and Alice's Story

In 1996, I worked with a woman named Alice. Sadly, as you'll discover as you read, Alice's husband's vulnerability to her constant health and unhappiness challenges resulted in the deterioration of his physical health. He was so concerned with helping Alice, finding the financial resources for her treatments, and working long overtime hours to support her that he never spent the time to check in with his health. For years, Alice had symptoms including insomnia, indigestion, headaches, and anxiety, to name a few. They were not understood or successfully treated by her regular physicians. No one could help her. After much expense, she began to search for answers outside the box. Alice and I agreed to work on her energetic imbalances during a four-week program. The program would allow her to identify the patterns that created the greatest dysfunction in her daily life. As we dove into her story, we began to put together the pieces.

Alice began to see that the life she had been living over the past nine years had created a sense of unhappiness. A few years into her marriage, she had noticed that she lacked the drive to cook, clean, and engage in sexual relations with her husband. She had loved doing those things before her marriage. At the beginning of her change of heart, James, her husband, did not mind. He remembered her fun-loving and carefree spirit. He figured it was just a phase. As time went by, James realized that he did not have a voice in the relationship and gradually allowed her to disconnect from him and her commitment to their partnership. Because she was the driving force in the marriage, he felt he had no choice other than to accept and submit to her withdrawal. Because of his childhood experience, always being the second fiddle behind his brothers and sisters, it was second nature to allow this behavior in his marriage.

Initially, James had signed up to be with the woman he adored and enjoyed. Together, they had planned many adventures for their future. Alice, in turn, signed up to be cared for, just as her mother had. However, James seemed to feel that there would be an equal give and take in the relationship. They had two very different ideas and intentions going into the marriage, which they never really discussed or acknowledged. Although James agreed to care for Alice financially, he did expect them to work as a team. He wanted Alice to participate in his life. James wanted them to engage in the same activities as before they had married. He wanted her to embrace hobbies, friendships, and organizations that were close to her convictions. Alice quickly became disconnected. She was not able to interact with any enthusiasm or joy. Alice started to feel incomplete. The distortion she felt about her life took on a life of its own.

As Alice and I worked together to create a new sense of awareness and empowerment within her, without her needing or praying that James would change, the energy of the relationship started to shift. She began to see her husband as a benefit instead of someone to ignore. Alice even started a routine of cooking meals when he returned from work so that they could communicate together once again after fifteen years! Through her new connection to awareness, Alice began to feel the love that James had always provided for her but that she had been unable to accept. She also started to feel stronger and more at home in her body.

After the four weeks were up and the program had come to a conclusion, Alice went off on her own to create a more intimate relationship with her husband. She found that cooking meals for him, cleaning his clothing, and waking up in the morning to greet him with affection made her happier and more in balance and at peace.

A few months later, I received a phone call from Alice. She communicated that James was sick. Over the years, she had gone from doctor to doctor, clinic to clinic, and had not paid any attention to his health. He was the provider. How could he be sick? He never got sick nor took time from work because of illness. Throughout the years, she never asked or knew if there were funds for her hospital hopping or her numerous outlets for healing modalities. She never asked him if he needed anything. Being so engulfed in her misery, she had not had nor taken the time to check in with him.

Upon the news of his illness, Alice suddenly became concerned about finances. Finances had never been a consideration while she spent thousands of dollars on "short-fix" cures with no long-term effects. However, now, today, during a routine examination to keep the family's health insurance active, the radiologist found a large tumor in her husband's chest. The biopsy came back as stage 4 lung cancer. Through further diagnostic testing, they also found tumors in his throat area (the communication energy center). The physicians urged him to check into the hospital immediately for treatment.

Alice mentioned to me that she recalled that James had complained of pain in his throat while breathing from time to time. Because she was so busy with her misery, Alice had not been able to hear her husband's call for help. He was in the middle of a toxic and unhealthy home environment and could not breathe. He was stifled and shut down. He tried over and over again to help Alice instead of himself. Neither Alice nor her husband listened to his needs. Within four months, James passed away.

For so many years Alice had drained her husband with her recurring challenges, with no successful long-term remedies or conclusion. The ongoing energy needed to support his wife's sicknesses, the financial responsibility, and lack of affection for years ultimately drained him of his life force. When we are unable to communicate our truth to others, especially those we love, stagnation and inflammation can create serious complications. Alice's husband endured years of unhealthy turmoil, drowning in the emotional and mental agony over the relationship with his wife.

Sadly, James did not develop his self-confidence or tools to support his communication energy over the years. Because of his childhood issues, he was not able to speak to Alice openly and honestly. He merely went on until he could no longer function. James passed away quickly after his diagnosis.

 Intuitive Energy Center

CONNECTING TO YOUR NATURAL SENSING OF THE TRUTH, ABILITY TO SEE BEYOND NORMALCY

Life Experiences or Feelings That Can Create Imbalances

- Knowing the truth and disregarding it
- Hearing the truth and discarding it
- The feeling you know something and closing the door to the knowledge
- Inability to communicate what you know to be right
- Failure to share your experience from your perspective
- Struggling with what you know and burying it within

- Speaking the truth and not being accepted or acknowledged

- Inability to focus and feel hope

- Failure to overcome depressive thoughts and feelings

- Migraine headaches caused by knowing and thinking with no resolve

- Physical imbalances caused by overthinking and discarding the truth

- Knowing the truth and being forced to pretend it is not the truth

Intuitive energy is a crucial energetic component to the magnificent system living within the human anatomy. If we sense the truth, are sent signals, and discard the information, life can seem more difficult than necessary. Intuition comes through three main areas of the human energy force: the head area, the stomach, and the heart. Our brains can receive signals that conflict with the data stored to alert the decision-making process. The heart knows the truth; a feeling in the heart will signal a sensation to "feel" your way through the moment, and the stomach, attached to the vagus nerve, will send precise signals. Our bodies and our spirits work in tandem to assist us in enjoying this life experience to the fullest possibilities.

Client Story

INTUITIVE ENERGY 2004

Patricia's Story

One of the saddest stories I endured as both a mother and professional intuitive was Patricia's story. Patricia was living on the East Coast and had a son living on the West Coast while he attended college. As a young adult, he had suffered from drug addiction. This story describes the incredible connection that Patricia had with her son through her intuition, her insightful self. Many parents seem to have this connection with their children. They know when they are sick, feeling sad, feeling lonely, and need assurance. Patricia and Adam share a particularly heart breaking story.

When Patricia and I discussed her health challenges, what I noticed immediately was the strength of the energy associated with the experience she had endured. The sensation that was affecting her physical body was intense and very real. I was nervous for her as she opened up while I accessed her heart energy. Her sense of attachment to her beautiful son was incredible. Every time her son needed help, she could feel it in her heart and her stomach. She knew it, and he knew it. It was almost as if they were still connected by the umbilical cord that had brought them together from the creation of his life.

It was not always a safe and comfortable feeling for her. When Patricia's son was in high school, he developed an addiction to heroin. This realization shook and collapsed the entire family. When he went through the challenges of drug addiction, Patricia herself developed difficulties in her stomach with ulcers (self-assurance) and high blood pressure (love energy). Each time he was able to overcome his illness (addiction) she too became a little stronger physically.

Adam had been clean and sober for over eighteen months when the family agreed that he would attend a university on the West Coast. He was feeling healthy, confident, and stable. He was ready to embark on a beautiful journey with other college students. The only person who would be left behind was his old high school sweetheart, Alecia. They had grown up together. She too had challenges with drug addiction. She was the person who had introduced Adam to heroin in the ninth grade.

They had not been together for over a year. Adam felt that he was strong enough to move away, be on his own, and become independent and educated—something that was very important to him. He wanted to prove to his mom that he could be successful, especially after everything they had gone through together. He wanted a fresh start, including new relationships and new surroundings.

The first and second semester went very well. Adam maintained As and Bs in his classes and started to truly enjoy the closeness of his new community and life away from his parents. He met new friends, was clean and sober, and began to feel good about his life and himself. Then Adam received a phone call from Alecia, who wanted to travel to the West Coast for a visit. Alecia asked Adam to keep the visit a secret from his parents and her parents. He agreed to keep it a secret.

The following day, Patricia called to check in on Adam. She told Adam that she felt he would get a message or phone call from Alecia. Patricia intuitively saw Alecia coming to visit and knew that it would not be a good idea. She had a terrible sensation in her stomach and heart when she thought about Alecia's traveling to see Adam. Adam was alone, trying to start a new life, and needed time to ground himself in the new environment. Adam did not want to break the promise he had given to Alecia, so he heard his mom and said he would take precautions if the situation arose.

Two days later, Alecia arrived in northern California. She was a wreck. Adam immediately knew that she was on heroin again. He felt compassion and sadness for her. He took Alecia to his apartment, and they talked for hours. It was Thursday evening, so they just hung out together. The following evening, Adam received an invitation to a massive gathering at the university, with bands, booths, alcohol, and lots of weekend activities. For some reason, Adam started to drink that Friday night and then he and Alecia started to party like old times. Little did Adam know that later that evening an event would be put into motion that would change his entire life. His mom tried calling him several times, but he put his phone on silent as he did not want her to know that he was drinking and certainly did not want her to know that Alecia was in Berkeley.

At 1:30 a.m., Patricia was in her bed sound asleep. She suddenly bolted up, filled with anxiety and fear, felt her heart rapidly beating, and saw the vision of her son calling out for help. Patricia remembered trying not to hyperventilate. She tried to relax as she lay back down. Patricia tried to get her balance while gathering her thoughts and feelings. She grabbed her phone and immediately called her son—no answer.

Within ten minutes, she received a call from one of Adam's roommates. Adam had just passed away from an overdose of heroin. They had called an ambulance when they discovered his lifeless body, but, sadly, he had already passed away. Alecia flew home the next day, back to her parents. The roommates disclosed to Patricia that Alecia had arrived in Berkley just a few days earlier, the day after Patricia had called Adam to warn him.

The day I received the call from Patricia, it had been ten years since the passing of her beloved son. The pain Patricia felt in her heart was as real and alive as if it were yesterday. The reason for her call was to gain information on how to settle and calm this vital energy as it was affecting her health in profound ways. As I mentioned earlier, this pain, this agony, this severe suffering had touched her heart and her mind, and it was destroying her health, even though it had occurred ten years earlier.

We worked together for several weeks. Patricia was able to obtain answers to her yearning questions surrounding the life her son had created on his own, as he was doing his very best to step into manhood. Answers, confirmation, and hope were the tools Patricia needed to start her slow recovery from the devastation she felt surrounding her son's very young life and tragic death; the death she was aware might take place as she desperately called to alert her son. The pain Patricia had been holding inside herself for not being able to save her beloved young boy almost took her own life. The blood pressure challenges, the anxiety, and the disconnection from the life she so desired to live all held nicely in an area that was creating severe emotional and physical difficulties.

Her story exemplifies that no matter how intuitive you are, you cannot change or help another person if he is unable to hear you, take your advice to heart, or take action to improve. The good news is that today Patricia dedicates herself to helping parents and kids who have addiction issues. She has genuinely used her life lesson to cultivate a deep connection to her intuition as a way to be of service to others.

Wisdom Energy Center

CONNECTION TO PERSONAL WISDOM FROM BIRTH, A KNOWING DEEP INSIDE

Life Experiences or Feelings That Can Create Imbalances

- Inability to speak the truth about what you know to be true

- Resisting your knowledge
- Being told the information you have or the words you speak are useless, without fact
- Being told you must keep your opinion to yourself
- Having insight and stuffing that inside your heart or brain without release
- Knowledge unseen and untaught being discarded.
- Inability to share your thoughts with those you admire
- Pushing your brain to accept information that does not feel right to you
- Being pressured to speak truth other than your own
- Being told that your knowledge has no value to others
- Sharing information with those who make you feel that you are in error
- Being told to keep your thoughts, feelings, and ideas to yourself

Wisdom energy arrives with us at birth. There are many things that we realize throughout our lifetime that we "just know." When they arrive in this world, children have wisdom beyond what we can believe. Only after being told time and time again that we do not know what we are talking about, do we begin the descent into losing touch with our information source. More and more, if you track the individuals who have made significant impacts in this world, you will find that they made decisions to leave the traditional academic track. The information they stored long before they entered society's path to intellligence, or the information they tapped into with their intuition and wisdom, far exceeds anything they could learn from a text or a professor.

Client Story

WISDOM ENERGY 2012

Bonnie's Story

A young woman came into my office just after graduation from high school. Her mother wanted to understand more about her nervous system disorder. Bonnie was okay under most circumstances, but at times her left arm and hand would tremble without cause. She had undergone MRIs, CAT scans, acupuncture, massage therapy, and other treatments without any clear indication of the core challenge.

This disorder had been a concern from age four. When Bonnie started middle school, it became progressively worse, and her mother started to search for answers. Years had gone by without any concrete understanding of what the actual cause was or what the successful long-term treatment might be.

During our first session together, I spoke to Bonnie about any concerns she had that would create a sense of uneasiness within herself. She did not feel that there was anything in her life that caused great upset. She enjoyed friendships, had a boyfriend she was pleased with, and looked forward to college.

My intuition showed me that from a very young age she had been involved in an environment that was driven by a female who was angry, repressed, and filled with hormonal challenges. I quickly identified this to be her mother. It was clear from our first meeting that her mother was not open to returning to that period of her life.

So I took a different approach. I pulled out the tools I love to utilize during my initial consultations. I wanted Bonnie to access her wisdom, regardless of where it was hiding. With her assistance, I knew she could come to understand what provoked this nervous system challenge. As I put the cards on the table, she quickly and excitedly wanted to know what the yellow and purple cards meant. Those cards represent Self-assurance and Wisdom Energy, I said. I smiled as I started with my simple yet complete explanation of the energy centers; she was excited to learn more.

As she started to talk about various things that challenged her self-assurance and self-esteem, her arm began to twitch. It was very apparent that the challenges were stuck in the spinal area behind the stomach, upward to the neck, and into the back of the neck, where the energy channels are all connected.

During the time we were together, she continuously spoke about her mother's anger. Her mother's behavior was something on which she could quickly become fixated. She could not understand why her mother could not find calm or happiness.

Bonnie's mother's anger, fear, and stress had a direct physical effect on her causing digestive issues and headaches. Her mother was still under so much distress that she was not able to always express her physical challenges. The fact that Bonnie repeatedly told me that she had not felt any distress over her mother's aggressive, angry behavior, while she held her stomach, was a situation I see often. A child does not want to admit that the unhealthy environment they were raised in caused the damage, because of their love for the parent.

From a young age, Bonnie learned to stuff her emotions, fears, and challenges behind a smile, because her mom felt she could not change her lifestyle, resulting in Bonnie's living in an environment that caused her distress. Bonnie felt that all areas of her existence were difficult. As Bonnie went through school, she always felt that everything was complicated, and nothing would be or could be smooth and beautiful. She described her life as difficult, stressful, without joy, painful, and hard most of the time, but that she had deep feelings of love for her mom. She did not blame her mom, nor did she think that her mom was directly involved in her physical/emotional challenge.

These loud noises that "seemingly" sounded like the sounds from her childhood triggered the neurons to signal danger. When this type of environment was present, Bonnie was not able to control her nervous system, which automatically went into stress mode. The remembrance of the emotional stress from her childhood affected the nerves running from her stomach to the heart, to the back, and down the arm channel.

Bonnie started to understand where the blockage, stagnation, and trigger lay. Bonnie's healing journey began the moment she started her journey to healing the wisdom and self-assurance energy centers that were damaged. With the help of shiatsu, acknowledgment, and staying away from environments that she knew would be disruptive, she was on the road to recovery and the strengthening of self.

She has also started to understand that even though her life with her mother did involve stress, anger, upset, and fear, it was not the life that she would create in her own world. Through this process of learning, her mother also learned that she, too, did not have to carry this burden that was no longer necessary. These two women loved each other. Because the energy was not intentional, and her mother had tried for years to help her find answers to her nerve challenges, there was no deception energy involved.

Three Energies that Destroy Hope, Happiness, and Health

CHAPTER 3

One of my most significant discoveries in 2007 was how the accumulation of mental and emotional energy associated with fear, despair, anger, loss, jealousy, and expectations destroyed specific areas of the physical body, primarily the abdomen, intestines, thyroid, heart, and colon. Story energy is one of the most difficult to shift, closely followed by expectation energy. When this aggressive type of mental and emotional charge travels through the physical body, various symptoms will begin to surface. If the symptoms are not addressed, through a deeper understanding of the culprit, the neurons can start to destroy the pathways necessary for a healthy balanced body.

Diseases directly impacted by Illusional Energy, Expectation Energy, and Story Energy:

- Immune System Disorders
- Some Cancers
- Blood Disorders
- Cardiovascular Imbalances
- Stroke and Bell's Palsy
- Nervous System Disorders
- Hormone Imbalances
- Bipolar Disorder
- Digestive Diseases

Illusional Energy

Being truthful with ourselves and others is imperative for growth, understanding, and creating what we truly want in our lives. Being honest with others, if they are involved somehow with our dreams,

intentions, and plans, is critical to our well being. What I call "illusional energy" is something humans can engage in because of lack of self-esteem or the intention to manipulate. It is indeed a waste of energy that permeates and contaminates all of the nine energy centers and can create imbalances or illness. With some, it seems to be an everyday fact of life.

I define illusional energy as energy or a thought that is created by the person who has the desire to accomplish a dream, goal, or fantasy but will not communicate the plan or desire to the participant or participants. It's a fantasy of what could be right and real—if! This lack of honesty does not allow the others to participate, interact, or engage in the experience. It leads to massive deception, manipulation, and ill will. Being out of integrity or congruity tears the physical and energetic body apart. Illusional behavior is hurtful and often so abusive that it can change the courses of people's lives in deeply damaging ways.

The energy utilized while creating illusions (illusional power) can be more than heartbreaking. It can consume large amounts of precious time in the lives of those engaged with the planning, thinking, and anticipation of dreams or intentions that have no foundation. The energy associated with this type of thinking or dreaming can be exhausting and can lead to illness, suffering, and mental and spiritual grief. Illusional energy is a disappointment (unrealistic, brutal) and it inevitably backfires because there is only one participant. Why is creating illusions so popular? Why do individuals create scenarios in their minds, their hearts, and hold them inside without sharing them with those participating in their fantasies? Fear! Low self-confidence. Inability to execute desires. Master manipulation. There are many reasons.

Human beings are so often run by fear and by not feeling good enough. Failure to admit the truth in many situations is due to the dismay of failure, judgment, or rejection, which may or may not be the actual reality. Without testing the water in the real world, the illusion begins. For this type of individual, the plotting, scheming, and manipulation is more comfortable than honest direct communication. All of this rational reasoning is created solely within the mind of the "planner" without the participant's knowledge. It is a kind of false advertising. The truth is that the dream which leads to disappointment has been created independently in the dreamer's mind before the interaction. By not allowing others to have access to the entire truth of the matter, precisely what the dreamer has created in his mind and heart, others can only respond to the seen or misleading communicative energy. They have no idea what the dreamer has created in his mind and spirit if he has not expressed or communicated his scheme or plan to them, in truth. Is it clear that the initiator is the person who has created a platform for failure? This person is flying solo. He or she is in a dialogue that basically, and sadly, does not exist, quite yet. Yes, individual pieces may be applicable, but the core is not. The core desire or fantasy, however, is left off the table. So there is the pain, suffering, confusion, and manipulation, leading to failure.

Manipulation without the other parties' knowledge is such a waste of energy, emotional engagement, and mental power. Regardless of how false the fantasy may be, many will hold on and believe that obtaining the goal, in the battle, will magically create the results they desire. So in the mind of the planner, the plan is made, there is a waiting period, risks calculated, and finally, almost always, the goal is not reached in its complete desire.

Alternatively, when the idea does work out, in some version, the feeling behind it is not what was desired or needed. Enormous amounts of energy are necessary for the plans made within, which almost always end with feelings of depletion, despair, and sadness. This energy scenario takes several parts of the brain to work excessively. Anger is an issue, but lies beneath the other feelings.

I have seen this cycle repeated over and over again. I have worked with many people, both young and old, male and female, who have created in their minds the scenarios they needed to keep their lives moving in the directions that had brought them hope, possibilities, and attraction to what they truly desired. Unfortunately, they had not told their partners, or the partners they wished to be with, the entire dreams or creations they wanted. After weeks, months, and, for many, years, the disappointment started to build energy all of its own. The more they strove to obtain their secret goals, the more intense and disastrous the energetic connection between them and the other person became. The results were generally horrific when one partner did not know what was going on, while the other had been carefully calculating his or her every move. The others involved in the scheme, participating unknowingly, had absolutely no way to defend themselves. They didn't understand why they almost always came up empty in the ability to satisfy the friendship or relationship. Often being caught off guard, they had no idea what the right response was from day to day. For many, relationships ended without a real understanding of what had gone wrong. Because the truth was never exposed, there was loss without an opinion.

Be mindful to speak the truth as often as possible. If you are not speaking the truth, not allowing the person you engage with to be able to have free choice to either participate with you or not participate with you, you are taking that which is not yours, their energy, time, and goodwill.

Engagement with others must require truth and respect if you plan to grow the friendship or relationship.

Do not waste valuable time on a dream or idea that includes another when you are purposely manipulating the situation. This behavior will not allow you the peace of mind or inner wellness required to obtain real goals.

Creating an illusion will hold a lot of emotional challenge and disappointment for you and the other person. When the illusion fails, the dream will play itself out over and over with frustration, anger, and sadness. Sadly, the person who was not privy to your illusion will never know what he or she has done to disappoint you or make you feel unhappy. Honesty with your expectations and desires will allow you and the other person to make decisions and choices based on truthful interaction.

How can the right mate, relationship, career, or emotional experience find you if you are not honest and forthright in your desires? Mixed messages are a waste of time and precious energy. When you communicate the truth about your wants and needs to yourself, your life will become more and more balanced, and the necessary alignments will appear.

The appropriate person or experience will often show up to accommodate your desires, so be truthful and forthright, and press on!

Expectation Energy

EXPECTATION DESTROYS LOVE, RELATIONSHIPS, AND INTEGRITY

There are many examples of expectation. There are many different ways we communicate expectation: we do it verbally, visually, in writing, through nonverbal expressions, and more.

During consultations in my office, I suggest that the participants jot down the expectations of their relationship, marriage, with their children, and with leaders in their workplaces. It can be quite a shock to those who are not aware of just how demanding and insensitive their expectations are and have been. It can create total collapse and destruction in others.

Have you ever wanted to take back or retrace what went wrong in a relationship? Each relationship challenge that I have worked with has had one common complaint: my expectation destroyed the relationship. The person was not able to communicate this truth initially. It took her looking inside herself to find out what had gone wrong—the unmet expectation that had led to the collapse of the relationship. Why do we feel that we have the right to demand so much from those in our lives? Where are the needs and self-righteous behavior coming from that allow expectation to gain ground in relationships? To belittle, control, and humiliate those who come to assist, love, and protect is not an honorable trait. Although it is almost always unintentional, the energy behind it is very constant. The power of expectation has a pulsation that is demeaning, uncaring, and strangling to those who interact within this platform, even though the person administering the expectations "seemingly" displays this behavior under the guise of love and support.

Not reaching the goals or demands set by the person or persons administering the expectation energy will weaken or potentially destroy the self-confidence energy needed to sustain a balanced life. More importantly, it causes deep pain in the heart, when the other person is always feeling a failure. When a child or children have not met parental expectations, time and time again, a child will become less than the person he or she is capable of or has the potential to be. In relationships, the deep love that once existed can and will diminish. Being beaten down by expectation destroys the spirit, especially when the engagement is with loved ones.

Over the years, I have worked with many women. In 2010 I created a study with the data of individuals who had worked with me from 2000 to 2010. The focus of my research was expectation energy that destroyed marriages. I used five hundred relationship examples to gain my data. The following information is directed to support women who are having difficulty in their relationships, partnerships, and marriages. If you are a woman who has had challenges in your marriage or relationships or have female children, this section may be of particular interest to you.

In many households the energy of the woman is at the forefront of the family's success; not all, but certainly a large percentage. A huge challenge I have faced numerous times during consultations is with the expectations women feel entitled to during their marriages. Women who need what they need and will not accept anything less can become completely imbalanced within their lives. They can lose connections to passion, love, and excitement. When demanded needs are not fulfilled by their partners, husbands, or children, the fury inside can grow. Without stepping back to review the patterns of this destructive behavior, the damage can be so severe that relationship respect can be lost forever.

Women have told me that they need men to be precisely what they want, and without their expectations being met, they use withholding as a means to gain control. This exchange can and does result in dangerous implications that can at times destroy the very foundation that must be solid for a family unit to survive and flourish. It is not until the marriage has failed and the intimacy has passed that a few of the women finally understood just how difficult it was to achieve all they desired from their mates.

If you are personally involved in this type of expectation energy, please review what is important in your life. Reflect on the power attached to these expectations. Realize your intention, realize the feelings you have for your spouse, lover, and/or children, and move forward according to your heart. Expectation will take the breath out of a relationship. I have worked with over a hundred women who developed breast cancer or cysts in the reproductive area of their bodies, all associated with the alignment of expectation energy. The neuron pathway that travels from the brain, to the heart, to the stomach, and through to the colon, transmitting expectation energy is dangerous for women.

Because of the amount of pressure we feel in life, no one has downtime. No one has a minute to breathe, so we shove all the responsibility we feel onto those closest to us. We forget that they too are experiencing that same feelings of being overwhelmed. We start to live in our little worlds of stress and unhappiness. Instead of seeing these people as assets and supports, we start to see them as the ones who either make or break our success. The battle begins.

Many men believe that if they can be better at being husbands, boyfriends, or partners, then everything will be ok. They believe that if they can make it happen, their wives, partners, or lovers will return to be the loves of their lives. When there is expectation energy, this is nearly impossible.

Men are not energetically wired to handle the energetic responsibilities of the female emotional energy. It is a difficult task for them, even though many truly want to.

Men want to make their woman happy. Unfortunately, almost always, they do not know how.

To understand the energy that motivates and destroys love, respect, and hope between men and women, there must first be an understanding of how each participant operates. This takes dedicated partners with strong motivation to succeed with each other.

Story Energy

THE ENERGY THAT REPEATEDLY HOLDS YOU BACK FROM ENJOYING LIFE

What are you feeding your heart, brain, vagus nerve, and neurons?

"Story" energy is energy from the past: pain, loss, sadness, anger, regret, disappointment, and despair—the what if, what could have been, and what was so much better than today. It becomes an energetic loop that is relived year after year, slowly killing any possibility of happiness. It comes from feeling a lack of closure or resolution related to traumas, deep pain, unresolved heartache from grief, and those life moments that caused a crippling feeling

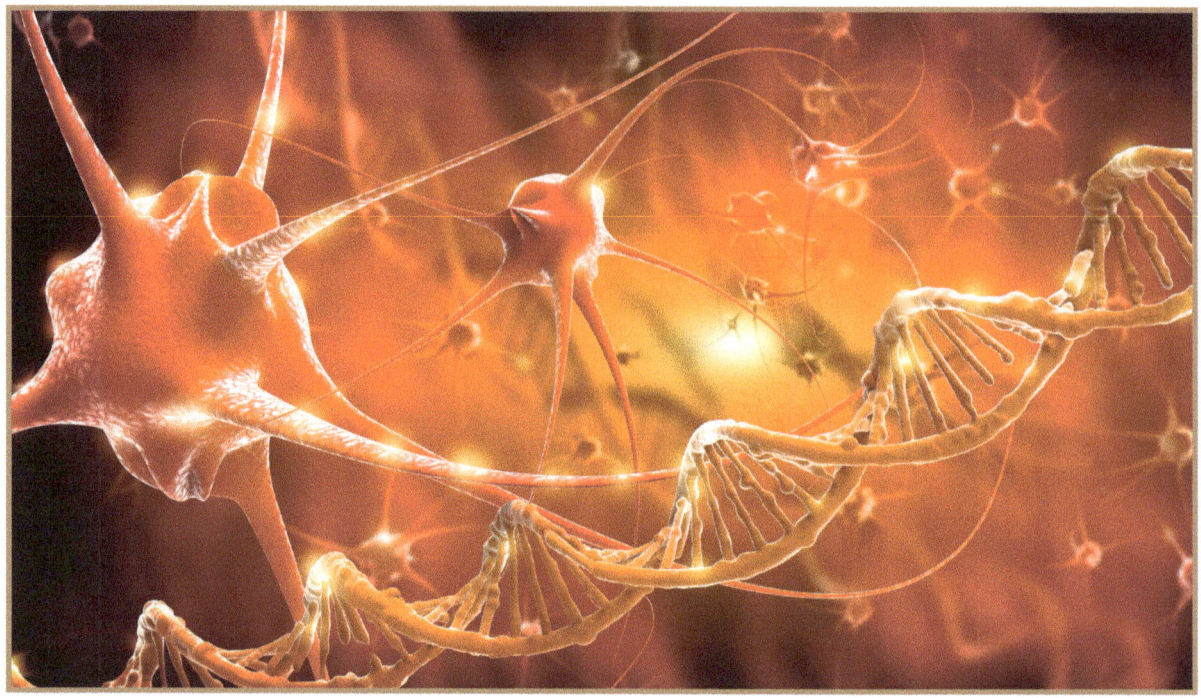

inside. Stories will haunt the spirit and those connected to us. How many times have you rehashed your story in your mind? How many times have your feelings affected your attitude, energy, and prospects for the future? How many times have you told your story and wondered, "Why do I keep repeating myself?"

Story energy can dominate your mind and can become an obsessive cycle that runs your life. If your story has no sense of resolution or conclusion, it can travel ruthlessly throughout your brain, gaining more and more aggressive power. Do you know anyone in your life who continuously tells a story? Have you ever wondered why a loved one could not quite get over the trauma or painful event from the past or why you cannot seem to get through to the person who is stuck? This challenge, which is more common than we have yet to realize in our world of medicine and science, is due to the power of the neurons and the vagus nerve.

Why We Repeat Story Energy

Without understanding the dynamic sensing system inside the human physical body, it can be challenging to move forward. The past can hold severe pain, significant loss, and fragile confusion for many people. If a pattern developed stemming from the past and is currently circulating through the neurons, vagus nerve, and other areas of the physical body, it would take a significant commitment to repattern this energy. There are many theories of how to move past the "past." Without the understanding of the neurons, vagus nerve, and the Invisible Anatomy's participation of repeated patterns, it may be difficult for some people to move forward with long-term success.

Story energy, if it includes unresolved pain, can be dangerous to the human spirit and human body. It takes away the energy we need to create passion, excitement, dreams, and possibilities for now and the future. If the story energy begins in the early morning, there can be a severe drop in the energy resources available throughout the day. Priorities will blur and the day will start with little enthusiasm.

If you are like most people under the excessive stress of today's life, you will need all the energy you can get to deal with what's ahead on a daily basis. Can you possibly access or utilize the power you need to operate on a regular basis with optimal energy, hope, enthusiasm, or optimism if you are also dealing with story energy from the past? The answer is no!

What part of your day's available energy has been used up by your mental chatter before you drink your first cup of coffee, water, smoothie, or tea?

Story Energy Creates Exhaustion

When you wake up first thing in the morning to the same energy that exhausted you the day before, the cycle can begin to take on a pattern. We create each day from what we feel first thing each morning. Our available daily energy is directly attached to what we think and feel. If the story energy begins to develop a pattern, the story energy gains momentum. After twenty-six to thirty-seven days, the story energy will begin to circulate throughout the body via the vagus nerve channel. After this time of repeated chatter and dialogue, the pattern is set, with or without your consent.

Please take a moment and define for yourself the story energy you hold or that holds you within your own daily life. What challenges do you feel occupy most of your mental and emotional space? What percentage of your day is unintentional, mentally or emotionally, devoted to relationship challenges, children challenges, financial worry, past pain, disappointment, despair, fear, anxiety, career uneasiness, or more?

By tracking this information, you can start to change the amount of time and energy you spend reviewing and reliving the past. Living in the past does not allow for embracing the future. Get to know the neurons, the vagus nerve, and the ability your incredible body has to assist you with adopting a new way of living your life.

Each Day You Awake – Make Choices that Connect You to Your Passion and Dreams

Let Your Spirit *Know* You Are On Board for This Life Adventure!

Parents and Caregivers:

Remember to remind your children that although one moment may have seemed like a failure, each day new opportunities arise that will assure us that the moments will pass that have made us feel "less than." Failure is not a word to be circulated in the human mind or heart.

Courageous Individuals
Hitting Life Head On

CHAPTER 4

Books can be tricky concerning hearsay and specific factual information, especially in the self-help arena. Sometimes, evidence provided from the author's point of view is the truth from his/her perspective. It's always been my desire to let individuals speak for themselves, without my thoughts and conclusions, regarding the steps they took in their journeys to wellness.

The individuals you will read about have chosen to take the path to healthy lives, regardless of the pain and uneasiness necessary to succeed. They are the real champions in my books. Let me introduce you to some of the favorites over the years. It's been my experience that people love to hear my personal story, but today, I am introducing you to the real-life heroes of this book.

My team and I asked the following of those telling their stories:

"Please tell us what you learned during your time with Jennifer."

We have included their photos. Together, we collectively agreed that seeing the people telling their stories would give greater empowerment to the experience and the feelings associated with them. The individuals on the following pages were living their lives with challenges. We hope you enjoy their stories and, quite possibly, find inspiration to use on your journey.

I hope you enjoy their stories of courage and dedication.

Chapter 4

MEET SUSAN

Entering into my thirties was not all that it was cracked up to be. As a professional athlete for over fifteen years, I had endured a few dozen injuries requiring medical procedures. And although the performance pressures from the age of eleven felt difficult at times, nothing could have prepared me for the emotional collapse I felt after my beloved dog passed away from cancer, my decision to divorce my husband, and a move to a city without any of my friends or support.

After a medical consultation that left me filled with despair, the diagnoses of chronic fatigue and infertility—the words that I would not have the opportunity to have children—were too much to process. I felt like a lost soul roaming around. That same day, somewhat miraculously, I met a woman who told me about someone she knew of, whose name was Jennifer. I immediately set out to make an appointment to talk with this person with the hope that she could shed some light on my situation. The day I met Jennifer, my life began to turn around. The first appointment with her somehow shifted my entire feelings of despair. Perhaps it was the hope she gave me, the intuitive power she holds, or the gentleness of her heart. What stuck in my mind while working with her is that Jennifer taught me to be in awareness daily, living in the moment of now, first and foremost. She helped me put all the pieces in place that were surrounding me so that I did not lump them all together as the entire picture I lived in, which was drowning me.

Jennifer taught me the meaning of "take time to smell the roses and enjoy all the beauty that surrounds you." She said, "Be the beautiful, intelligent, independent woman you are, regardless of what you have recently endured." Although these are simple words and phrases, somehow when Jennifer communicates them, they hold powerful energy.

She worked with me on the challenge with my hormones and the direction of my career, and taught me about the energy centers and how they affect our entire world. I was unaware that everything that surrounds us, our work, the people in our environment, and our choices, all have a significant impact on our well being. Growing up on the East Coast of the United States, being in the world of competitive sports, this is not subject matter spoken of that often. Because of the number of injuries I had endured as an athlete, that was the first stop in my work with Jennifer. Immediately following was the hormone imbalance and the chronic fatigue challenge.

She did not believe for a second that I would not have the opportunity to have a child, as my physician had told me. She did not speak negatively about the medical diagnosis; she just disagreed, quite matter of factly. The news of never being able to bring a child into my world was devastating. I wanted to believe with all my heart what Jennifer was saying. As I began the work ahead, following the

outline I was given to move into a stable place in my life and health, I saw rapid results. I learned how to create the experience I wanted and not live from a place of energy from my past.

Although Jennifer did not like to accept responsibility for my progress, she was the key to my understanding my spirit and connecting to the energy I have inside myself to make changes and create opportunities in my life. Her modesty did not allow her to take credit for the new life I had created, regardless of the pain and overwhelming sadness I had endured. She reminded me often that it was my strength, my spirit power, and my desire to create the world I wanted to live in that had brought my success.

My time with Jennifer has left me with two things that will always be a reminder of intuition and the power to know the truth: her vision to see that I would be working with horses in my near future, and that I could very quickly bring a child into my life. Jennifer could never have known that my dream had been to work with horses since I was a young girl. It was my dream job. And the moment she told me that I would have a child if I decided to changed my life, my perception of energy, and my belief that intuition is available to us all. The look on her face when she said these things to me with a matter-of-fact smile and confidence is ever etched in my heart.

—Mommy, Professional Athlete,
Elite Equestrian Trainer, California, USA

MEET ANNA

I met Jennifer Kaye in 2010, and I immediately recognized I was in the presence of an incredibly gifted healer and teacher.

What I treasured the most was her simplistic approach to the seven energy centers, which I fully understand because of my extensive background as an international yoga instructor. Traveling the globe, teaching workshops, creating retreats, and engaging in one-on-one informative healing sessions, I have met with many excellent teachers and healers.

Jennifer's approach is unique. Her understanding of energy and how it affects us was easily applied in my personal life and shared with others.

The Illusional Energy, Expectation Energy, and Love Energy resonated with my soul! I have grown from listening to my own energy and my own body's "feelings." My intuition became stronger as I was able to tap into my own natural, intuitive strength. Opening up to my own energy has allowed me to more smoothly read into the physical and emotional challenges of others.

As I share Jennifer's Energy of Life System work throughout Europe, I am able to see the difference in those who seek to connect to the energy of themselves.

I am eternally thankful to Jennifer and her work.

—Anna Inferrera, Milan, Italy

MEET STEPHEN

When I met Jennifer, I was not seeking direction or help. I believed that it was by chance. But I know now that it was not by chance that I was able to sit down with Jennifer. I was in the United States on a soccer coaching assignment, and was not expecting to encounter a life-changing moment. As I know now, most things in life are not by chance.

During my first unofficial meeting with Jennifer, she began to explain several things regarding how I had felt inside, but did not know why or how to apply the sensations and thoughts I had running through my mind. After listening to Jennifer's opinion regarding various feelings that I have experienced over the past few years, many things just started to click.

At a young age, I knew that I was entirely different from those who surrounded me. I could sort things out differently than others. Even though I did not fully understand the implications of what I felt and knew, I understood it was a natural sense. I recently learned this was a strong sense of empathy with a keen sense of insight. I have used both during my several years of my teaching and coaching to guide my students.

In my time with Jennifer, I can confidently say I know now that the energy in my life, my perception of the reality surrounding me, and my reaction to the interactions around me have all affected me. I have learned, through processing the information given to me by Jennifer, that I have a choice in what I get involved in. I have a choice with every opportunity and experience that comes my way. Listening to my inner voice is now quite helpful. It is a tool that has become automatic for me.

Truth be known, I had no idea what energy was or really meant, nor the effect it has on us. Although it made perfect sense, her words of wisdom and intellect are not information or theories you hear every day. But from watching my interactions with others or sitting for a moment when I feel a sensation, it now makes perfect sense. I am clearly aware of much more than I had been before. In England, we are not exposed to most of the self-help jargon that circulates in the United States. It's not a part of our general makeup of options. Stiff upper lip is our way of life. Jennifer is a no-nonsense woman. Her words are delivered with intensity and fact-knowing. It's hard to not listen when she begins to speak.

I have learned that when I experience feelings in my stomach, which I've had since an early age, it is my own self allowing me to know the truth about how I feel. My stomach is a focal point for me now. I understand that I can rely on the feelings that I have. In Jennifer's words, "Self-assurance Energy."

Since the time we met almost five years ago, I have graduated with a master's degree, taken a post as a school teacher, and have purchased a new home. With a great deal of passion, I report that I married my dream girl, Sara, in 2017.

I'm looking forward to taking all that I have learned about my personal energy into my future, benefiting both my personal and professional relationships.

—Stephen Potter
MSc Sport Coaching
PGCE Primary Education
UEFA B License, England

MEET TOM

My experience with Jennifer is as follows: Firstly, I encountered life.

When I began working with Jennifer, I had no idea that I would be diagnosed with cancer, that I would undergo radical surgery (and still have cancer), that a deeply loved family member and extended family members would also either be diagnosed with cancer or soon fall dead, unceremoniously and alone. I had no idea that my partner would be traumatized by severe false allegations that would tear his family apart. He and I had no idea that our lives would be forever changed by this complicated and elegantly swift thing called . . . life. All of this occurred in the space of one year.

To say that I overcame these overwhelming factors while working with Jennifer would be a vast understatement. However, to say that I encountered a bright, focused, unflinching, loving, tough coach, clairvoyant, and colleague would be spot on!

Did I mention my fears? My strengths? My doubts? My cynicism? My hostilities? My creativity? My resilience? Growth and love?

Not only did I encounter all of these, but I rediscovered the broader spectrum of "me at this moment." It's not easy to hear the truth she dishes out. As a matter of fact, it's downright unnerving at times. But, when you sit with her, you know she is telling you the truth from a very deep well of knowledge.

It was as if Jennifer accessed a bottle of intuitive window cleaner, sprayed it into the great panes of my soul, and began wiping away the decades of grime that prevented others from seeing in and my seeing out.

I believe that I have greater clarity now, the tools to maintain my personal upkeep, and a better sense of the vast possibilities of my life.

Relearning that human energy is a limited and precious commodity not only resonated with my training as an analytically-informed psychotherapist (see Freud), but it also helped me build more mature respect and humility for the mind, body, and emotions.

And finally, I learned that trust goes a long way.

—Tom Price, Licensed Psychotherapist,
Case Manager, State of Washington

MEET MARK

I met Jennifer several years ago. I was introduced to her by a personal friend who knew about her from a well-known attorney at a prestigious firm in downtown Seattle.

She was introduced to me as an intuitive, a person with a gift of knowing things. I had no idea what that meant exactly. I am a black-and-white kind of person, with a particular interest in facts and figures. Upon meeting her, I was surprised to learn that Jennifer looked very professional and normal.

What I learned from my time with Jennifer is that intuition is a potent tool. From a young age, I knew that I knew things, without having evidence-based support, as things just came to me naturally. After learning about the sensations associated with intuition, I was at ease with my knowing.

Jennifer is kind, leads with her heart, is supportive in the process of teaching, and stands firm in her abilities as an extraordinarily gifted intuitive person. It's astonishing really. She knows things instantaneously, within seconds, that a person could not possibly know without using her inherent skill. Even with the gift, it's what you would call miraculous.

At lunch with my sister the other day, I mentioned to her that my intuition had become so strong and powerful that I had to stay in constant awareness of my gift when negotiating in my business. I mentioned my unfair advantage. Intuition, the extreme intellect without boundaries, can be a discriminatory tool, and yet hard to disclose to others. The fact that what I knew while in meetings did not allow a fair playing ground at times caused me discomfort. Intuition is an advantage that, unless you're using yours, is hard to explain to someone else. Keeping your integrity in check while living with and using intuition is mandatory to keep your side of the street clean and tidy.

The power of intuition comes with a certain level of responsibility. Because it gives you the upper hand, you are very aware of the uneven playing field.

—Mark Gordon, Founder
  Metro Structure
  Sound Development and Business Projects

Chapter 4

MEET SHARON

I met Jennifer without knowing about her incredible gift. Our daughters were best friends at school. I can't really remember the reason we met initially, although I will say that from the moment I met Jennifer, my life changed in ways I would never have imagined nor believed. It's not easy for a "normal" person to even comprehend the gift she has. It is truly incomprehensible. Jennifer's gift holds such responsibility. It is through her heart that she leads all that she encounters.

We met for dinner at a local restaurant in Bellevue. Halfway through the meal, she asked me without any notice, "How do you feel about your dad?" I was more than surprised at her question and was immediately on the defense. I didn't know about her gift, or why in the world she would ask me this question so completely out of the blue. I muttered something like, "We don't get along," and thought the random yet emotional questioning was over.

After we finished our meal she said very gently, "Rethink your relationship with your father; he's a wonderful man." I really didn't know what to say to her. I was put off and answered immediately, "You don't know my father or me." And she said, "Please reconsider your relationship with your father." Her facial description did not flinch or change. She walked away with a genuine, gentle smile.

As time went by, I learned many things from Jennifer. One of the most critical and life-changing events that occurred over the years was my relationship with my dad, Samuel Steinberg. During childhood, I believe many of us can have ideas and feelings about our parents that never really leave our minds or our hearts. We aren't always able to really know our parents as adults or humans; we only see the heartache and pain we feel as children, always wanting them to be different or see us differently, to provide more support as we see fit when we are young.

Jennifer gave me the blessing of opening the door to knowing my dad. And, as she had said, my dad was an extraordinary and loving man. Without knowing Jennifer and her gentle-direct (almost scary at times insight) approach the day we met, I am quite confident I would have missed out on one of the most significant gifts of my life, knowing my dad as the man he was and not the 'dad' I remember him to be. My relationship with my dad had been estranged for almost four decades.

My father was sent to an Auschwitz concentration camp, along with his mom, dad, sister, and brother. He watched as his mother and sister were taken away, never to be seen again, and his father and brother were worked to death. Literally, my father (a young man at the age of sixteen enduring the loss of his family, his country, and all he knew) embarked on the most incredible journey a human being can create. I didn't understand what it all meant as a child or an adult. The history of the holocaust is unfathomable. My father, at a tender age, was forced to witness things that I cannot comprehend, on any level

The dedication plaque reads:

Never tolerate bigotry
Be as generous as possible
Learn from the past

of consciousness. And yet, my father raised a family of four children and ran a successful business, with passion and unlimited energy.

Jennifer explained that each of us has a spirit energy that is connected to our own personal wisdom and insight. My father never looked back. He knew who he was, he knew what he was capable of, as a survivor left without family or friends, and dug deep to identify his strengths. As I grew into an adult, I never stopped to wonder about my father's childhood. I was so focused on myself and my youth, that I did not have the time or need to learn about his. Facing that truth at an older age crippled me for a time. He was loved by so many in his life and, sadly, I could just remember the days that we didn't see eye to eye, as happens with most parents and children. I had never stopped my own life to take the time or effort to explore my dad or what he was about, until I met Jennifer.

My father passed away in 2010. Through Jennifer's encouragement and love, I was able to spend the last years of his life by his side as the daughter I had always wanted to be, but didn't honestly know how.

Weeks before his death, my dad reached out to Jennifer. Although I am not aware of the context of the conversations, I know they were both honored by the communication. My dad was not a man who believed in thinking outside the box. When I introduced Jennifer to my dad during a conversation, he knew he wanted to speak to her. There is something about Jennifer that is so honest and kind.

Thank you, Jennifer! Your gift still amazes me after all these years. If you only knew how it affects others. It's my biggest hope for you, as you travel through your life, to see the impact you have on others in the most profound way imaginable. I hope you know this before you leave your life. You are a godsend, and yet, you really have no idea.

You can still see my father's hard work and financial success at work today through projects such as the Companion Care Program at UCLA Medical Center in Los Angeles, which provides specially trained volunteers who offer individualized companionship to older adult patients while they are hospitalized, along with other programs at UCLA Medical Center.

The Samuel Steinberg Foundation provides funding for the Gindi Maimonides Academy in Los Angeles, which honors orthodox Judaism and houses a holocaust museum on campus.

The last portrait with Samuel's family, taken weeks before the family was transported to Auschwitz concentration camp. He was the only member of his family to have survived the camp. He kept the picture until the end of his life.

> *"When you are aware of wrongdoing to others, do something about it."*

—Samuel Steinberg

Sharon Steinberg Kerr
Maine, USA

Chapter 4

MEET LAURA

After hours of unexplained pain in my stomach, I was rushed to the ER. Even with all the testing available to the physicians, there was no explanation for the severe pain I felt. We honestly had no idea if I was dying or not. There was no answer to my severe pain.

My daughter knew of a woman with a special gift who could analyze the root cause of health issues and said she would leave the room and give her a call. Within a few minutes, my daughter came back to my room at the hospital and told me that Jennifer had communicated to her that she immediately sensed an issue with my pancreas. Within sixty seconds, she told my daughter that the physicians could not see the exact reason yet, but within six weeks the condition would be apparent. I was shocked and bewildered that Jennifer could "see" the problem and pinpoint the exact area of concern within minutes, without seeing me in person or knowing anything about me.

After spending the night at the hospital while undergoing more testing, I started to think about what my daughter had told me. I set up an appointment with my general doctor and told her what Jennifer had said. We decided to take precautions and look at how to support the pancreas while undergoing over a dozen more tests.

Six weeks after my visit to the emergency room (as Jennifer had communicated) my husband and I were on a cruise in the Pacific Northwest. The pain returned suddenly without notice while on the cruise. I was urged to leave the ship to arrive at a hospital in Canada. I hesitated about going to the hospital outside of my insurance coverage. It's an odd thing that comes to the mind, even though I could not move and my body was in shock with the severe pain.

Because the information had been given to me earlier by my daughter, and later by Jennifer in person, I was able to provide the information to the ship doctor. He ordered a blood test, and within an hour, he found that I was suffering from a severe pancreatic disorder. I spent two days in the medical unit on the ship. Because I was able to tell the doctor what I had learned from Jennifer, he could quickly diagnose the situation and treat me for my exact problem. We saved time and then the money that would be needed to be airlifted to a hospital in Canada. I was so grateful to Jennifer.

I was able to ease into knowing that the road ahead had a blueprint. There was no more uncertainty looming around me. No need for further testing or poking around my physical body.

In the meantime, I had contacted Jennifer personally. Although it was so difficult for me to even comprehend a person could have this gift, I believed she did because she had told my daughter the exact problem over a month earlier, even down to the six-week future event that would occur, giving me confirmation of the pancreatic challenge.

Jennifer worked with me for several weeks. During this time, I was able to understand what was happening within my internal body that created the challenge with my physical body. The direct link was so hard to know because I had never been exposed to this type of thinking. Her wisdom helped me gain a new perspective on life and my relationship to those within my life, primarily my husband of thirty years, my daughter, and my son.

I had internal work ahead. It was not easy. I had created areas of my life that were not healthy for me and my family, whom I love dearly. The good news: I had an overview of how I contributed to my physical health condition and had the outline of what was necessary to bring my life into balance.

Jennifer comes to the table with a very candid approach. Even though it's not always easy to hear her words, she leads with her heart, and you know it when she speaks to you. She does not judge nor does she tell you what you need to do. All decisions are your own.

The most important thing I took away from my time with Jennifer is that my choices in my life have a direct effect on the overall feeling in my life. She is not interested in talking about how great her gift is, she is humbled by her gift and set out to encourage *me* to know my personal power and how "great" I am.

My sincere gratitude goes out to Jennifer. I hope she continues to use her unique gift and open heart to help others find their potential and personal power.

—Laura Vivolo, Mother, Wife, Daughter,
Grandmother, Owner of Grandview Havanese

MEET ALLISON

I began working with Jennifer Kaye during one of the most difficult times in my life. I was grieving the death of my mother and the end of a long term love relationship. I was in so much pain. Daily sadness, confusion, and feeling stuck kept me from moving forward with my life. With Ms. Kaye's Energy of Life program, her gift of intuition, and dedication to my healing, my life changed for the better. I was able to ground back into myself, get clarity on my purpose, and use my intuition to guide myself back on my path. I'm grateful for Ms. Kaye's commitment to me living my life's purpose as a writer, keynote speaker, and advocate for women to be their true selves without losing their spirit.

—Allison Clay
USA - Europe
allisonclay.com

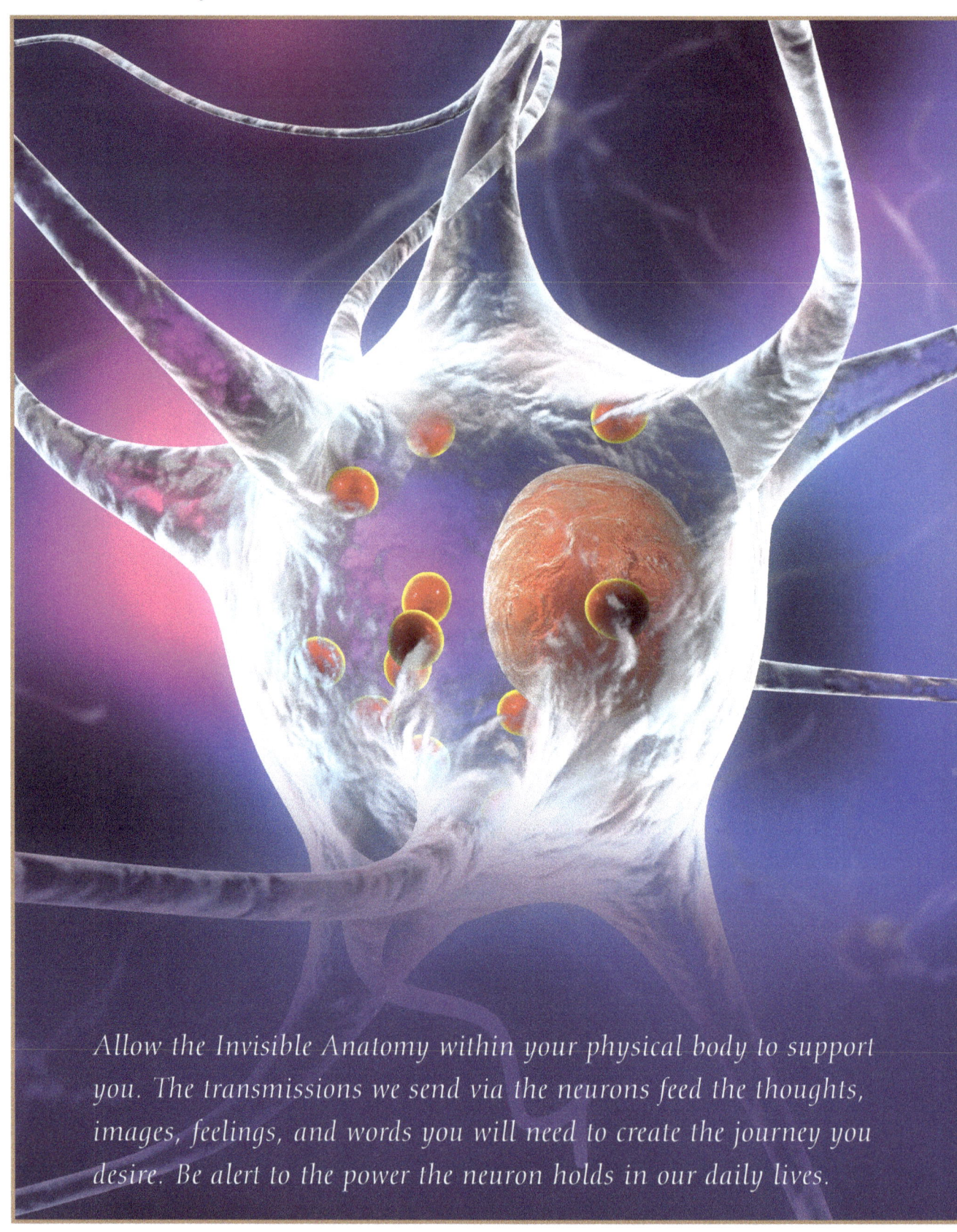

Allow the Invisible Anatomy within your physical body to support you. The transmissions we send via the neurons feed the thoughts, images, feelings, and words you will need to create the journey you desire. Be alert to the power the neuron holds in our daily lives.

EPILOGUE
KNOW YOUR TRUTH
PROGRAM YOUR NEURONS, KNOW THE VAGUS NERVE, LISTEN TO YOUR HEART

Because of my skepticism about intuition early in my life, it was not without a fight that I decided to help others with my gift. I was quite happy in the corporate world negotiating purposeful six-figure deals while enjoying my life. Creating success was always very simple for me. I did not feel the need to expose my secret weapon, to tell anyone about my intuition.

> The possibilities are unlimited if you allow your thoughts, your feelings, your infinite wisdom, and your human spirit to walk with you on your human life experience.

When I look back over the entirety of this experience—my journey using my gift—I must say that it is a great honor to have been of assistance to so many people who looked for answers outside the box. They dared to know within their hearts and minds that intuition had value and validity. Also, they believed in my gift. Together, we created a stable platform for learning, growing, and understanding the power of intuition. Without the thousands of opportunities with clients, research teams, and professionals to use my gift to dig deep into the dimension of "unseen," my life would have been very different. Although awkward in the beginning years, today I can say that I am very grateful for the time I've worked using my intuition.

Epilogue

As I discovered a relationship between the Invisible Anatomy, the hidden neurons, constitutions of blood, and the nervous system intricacies, it is with honor that I submit the data I have saved for over two decades. I hope that my work will continue to shed light on the imbalances in the physical, emotional, mental, and spiritual energy based on our life experiences. Without the love and respect I held for my mother's work, my perseverance, and the help of those who sought my gift, this would not have been possible.

I have had the honor and pleasure of working with some extraordinary people in my lifetime. I've met with individuals from around the world with diverse backgrounds, as well as ideologies that vary. One thing I have enjoyed about the human population is the spirit that lives within the human body. The brilliant intellect of those who live their dreams, live their intentions, connect to their purpose, and live with their knowing front and center.

I want to introduce you to information about various inventors between the ages of twelve to twenty. The more information you gather from the thinking of others, the less room is left for you to explore your wisdom, your intelligence. Below I have chosen a few of the inventors under the age of twenty who have created pretty cool things. Age is not a factor in *knowing* something is real or that you can create and build things with your passion and intellect, not education. The door is open for anyone and everyone, from all walks of life.

The possibilities for you are unlimited if you allow your thoughts, your feelings, your infinite wisdom, and your human spirit to walk with you in your human life experience.

Young Inventors

- Philo Farnsworth – Electronic pieces for the first television
- Blaise Pascal – Mechanical calculator
- Louis Braille – Reading and writing for the blind
- Alexander Graham Bell – Modern telephone
- Chester Greenwood – Ear mufflers
- Horatio Adams – Gum
- Frank Epperson – Popsicle
- Becky Schroeder – Glow-in-the-dark material
- George Nissen – Trampoline
- George Westinghouse – Compression Equipment
- Igor Ivanovich Sikorsky – Helicopter
- Walter Lines – Scooter
- Samuel Colt – Firearm
- Joseph Bombardier – Snowmobile

Leonardo da Vinci, Albert Einstein, Michelangelo di Ludovico Buonarroti Simoni aka Michelangelo, Steve Jobs, Walt Disney, and Henry Ford, help us understand that the dreamers and visionaries can provide extraordinary insights from a sense of what they know, without being taught. A unique polymath, Mr. da Vinci relied on his instincts and visions to create phenomenal world treasures in the areas of inventing, painting, sculpting, architecture, music, science, math, engineering, geology, anatomy, cartography, botany, and writing. Yes, he studied, but his extraordinary intellect was not learned. Where do you feel Leonardo gathered all that passion? His neurons were pulsating visions and intense focus, and although heavily commissioned (mandatory) during his life, he used personal energy to create masterpieces. Regardless of his life circumstances, being born out of wedlock to a peasant woman, a poor upbringing, Mr. da Vinci has been called the universal genius. He would certainly not fit into the cookie-cutter normalcy defined by today's society. How would "they" dissect this man in today's world of critical judgment?

Using your intuition will allow you a direct connection to unlimited energy. People have often asked me, "Where do you get all that energy?" When you let your brain rest, giving it time off, the entire body starts to come alive. I've been using my intuition solely, without my brain data, for over twenty years. Access to infinite energy is available without any effort if you open the door.

Open your mind and your heart to your intuition! When you feel "things," confirm your words, visions, and thoughts ASAP. Repeat the exercise of confirmation as often as you sense the signals. Give your brain rest as often as possible. I'm sure it's filled up with data. Open your heart to the senses you feel, giving your heart a chance to bring you the joy, happiness, love, and companionship you deserve.

BE COURAGEOUS! THINK OUTSIDE THE BOX! Your spirit knows everything you want and needs to fully experience this life with joy, happiness, and love, regardless of the trauma, sadness, frustration, and disappointment life gives us. Listen to your gut feeling!

LIFE IS NOT TO BE FIXED OR AVOIDED. LIFE IS NOT A RACE. Do not judge yourself harshly, nor remind yourself of the mistakes you believe you have made. Love yourself. Love those in your life with vigor!

Do not allow anyone to take your energy, wisdom, and personal intuition from you. Guard your spirit, your soul, with your life, literally. Your spirit does know everything you want and need to experience this life with joy, happiness, and love.

WHEN YOU DON'T FEEL WELL, STOP. LET YOUR BODY AND YOUR MIND REST. If you make yourself a priority in your life, you will find an unbelievable amount of energy available to you.

I hope you've learned a great deal about the Invisible Anatomy, the vagus nerve, the neurons, the energy centers, and INTUITION. Please take a moment to think about how long you've carried energy from life events or moments that you endured which took your passion, enthusiasm, and dreams away from you. Do not let life pass you by without discovering who you are and using your life energy to be the best you can be with your agenda. You are in life. You have a physical body. One thing I am very clear on, you have a spirit.

Don't allow anyone to strip away the time your soul, your spirit, and your heart need to heal from your personal life experiences.

Don't allow anyone to tell you that he or she has better answers to your problems than you have within yourself.

Don't allow anyone to take your precious energy who does not deserve it, and finally, DREAM BIG!

NORMAL *can be a hindrance to those with exceptional qualities.*

J.K. DICKINSON

Traits of a Nobel Laureate

He did not start to talk "on time," which concerned and agitated his parents. A physician was first consulted to understand what his problem was, and then a tutor was enlisted to help him prepare for entrance to a primary school. His slow verbal development caused concern to his parents and those around him and, yet, he was happy to take in all the sensory energy around him. He was labeled cocky and rebellious. He was expelled several times before he finally left school without a degree. This decision made it rather difficult for him to re-enter the academic world again. He demanded music in his world, socialized well with those he cared for, balanced a family with his passion for physics and mathematics, and communicated very well. He also fought authority and conventional thinking. He was not "a good old boy!"

This person thought and created in visual images and pictures, not words. The neurons running through his body flowed consistently through the energy associated with enthusiasm, passion, and wisdom. He spent a great deal of time sensing his visions and connection to his understanding without any assistance. The world at large still has the question: How did he accomplish his work alone? Words were not necessary and would not contaminate his neuron flow. This person was not open to opposing opinions, negative thinking, or assessment of his knowledge. He was passionate about what he knew and forged forward to compile the data. He only knew the truth, without hesitation or thought.

Who is this person?

Albert Einstein

The Lesson

What can we learn from the childhood of this Nobel laureate? "Normal" can be a hindrance to those with exceptional qualities. This type of story is in the same thread as many of the most extraordinary human beings who have hit this planet. They have one thing in common: they know what they know, and do not doubt their knowledge or vibe. The challenge these extraordinary individuals face at the end of the day is the judgment by those less themselves, who want them to "fit in."

Know your truth. Know what you know and do not allow the institution of WHAT'S RIGHT and WHAT'S WRONG to dictate how you interact in your world. Own your LIFE ENERGY!

The "system" dictating how, why, and when a human being is not normal is certainly not an accurate tool to access exceptional human traits, gifted individuals, and those spirits who are here to make contributions to society or simply to live life without the need to participate as soldiers in a line answering to all the commands.

Special Profile in Courage

Melissa "Echo" Greenlee

I came to Jennifer Kaye when I was feeling unclear about some important life decisions that were affecting my immediate future. Jennifer worked with me over the course of several months to discover my personal truth while also recharging my confidence. Working with her brought an enormous amount of clarity and personal power to my life. Before, I lived in a shell and liked to play it safe by limiting my own potential. After my work with Jennifer, a huge shift occurred in my life. I discovered the inner power of my own intuition emerge. I always knew I had sensations relating to truth in my thoughts and feelings and while in the presence of Jennifer, the door opened. Stepping through the door with courage and determination, I knew the truth I felt inside me was real and it doesn't require you to have functioning ears to hear it!

Being Deaf in an audiocentric world is not always easy. There is constant noise around you that you can't decipher. The communication barriers present in my everyday life are exhausting. Being a mother and differently-abled could have been challenging if I was not able to comfortably include my hearing daughter into both Deaf and hearing worlds. I wanted to share my passion and commitment to the Deaf community around the world by supporting awareness and growth. With my intuition and my personal wisdom, I have forged forward to make a difference in the world. Any obstacle can be overcome if you choose to dig deep inside yourself and connect to your spirit and personal power.

In my world of silence, the only thing I truly know I can hear is my own intuition. It doesn't require I have functioning ears to hear it. If that is all I truly can hear, I'm ok with that. Because with it, I've received the best gift of all. Total knowing.

 —Melissa Echo Greenlee
  Deaffriendly, Founder CEO
  Seattle, Washington

ACKNOWLEDGMENTS

First, I would like to thank my beautiful children. Being responsible for another human life has helped me grow in more ways than I could have ever imagined. Believe me when I say, "I am a big dreamer!" The journey with these two human spirits—Margaret and Matthew—has deepened my unconscious and conscious comprehension of life and the many rivers that flow into the enormity of the ocean. The love, the pain, the decisions, the anger, the frustration, the emotional and mental challenges, all of it, was never something I could ever have believed I would handle in the various ways I have. We, as a team, maneuvered through our growth. These two people are genuinely remarkable human spirits. Thank you, God! Thank you Margaret and Matthew (Maggie and Matt).

I know without any doubt in my mind or heart that my mother was the cornerstone for me to be the parent I am today. Yes, my dad, Dwain, was an enormous figure in my life, having reminded me of three essential virtues: don't cheat, don't steal, don't lie! However, my mom's energy was mind-blowing. Months before her passing, she was labeling, packing, and shipping all the items in her home that she wanted her family to have, with two tanks of oxygen in tote. My mom was one badass woman.

To the cool chicks in my life who have encouraged me the past few years, thank you! These women demonstrate integrity, love, honesty, and the will to continue regardless of what shows up in life. The love and support I exchange with these women have created a lifelong bond. I know, rain or shine, we show up for each other. Big sincere thank yous to my daughter Maggie, Lis, Lanie, Stephanie, Ms. Tina, Ms. S, Dianne and JDog.

A big thank you to Lisa Vivolo!

Sharon Kerr, thank you for urging me to write about my life's work.

To my sister Veronica, thank you. May God continue to bring happiness to you and your babies.

To my brother WD, you are the epitome of a genius. Your contribution to society in this lifetime will far exceed anything you are planning.

Mr. Potter, thank you, young man. The day you took me on the twenty-minute drive, arriving at the base of Mt. Rainier three hours later, was the day I emerged from a reclusive lifestyle. Stephen, I will never forget that drive. Thank you!

Dr. Janis Gruska, for always believing in me and my gift.

Also, finally, KL, for the unexpected phone calls right on cue while I mourned the loss of my mom in Italy, the support in helping me realize that the "show must go on" with my book, and so on. I know you rarely remember our moments, but I certainly do. The moments I've shared with you will never be forgotten. You've impacted my life. You've been a significant influence in my life, and I appreciate all our time together. You are one of the most intuitive on the planet! Take notice of my using initials instead of your "real" name! Thank you!

ABOUT THE AUTHOR

J ENNIFER HAD A DIFFICULT TIME admitting to herself that she had intuition and was reluctant to share it until later in her life. Thriving in the corporate world, starting her first company at the age of twenty-four, she lived a life with comfort, ease, and privacy. In her own words:

"Embarking on this journey to help others, by using my natural gift of intuition, initially brought great uncertainty. Telling people I could "see things" and knew the "truth" sounded far-fetched—even to me. I remember how uncomfortable I was watching my mother use her intuition to help law enforcement.

Due to my reluctance to admit that I had a gift, it took me several years to share my intuition. I previously had a successful corporate career—one that I used to consider my "real" job.

"My intuitive work started by working with physicians. I knew the field of medicine was the ideal industry to bring credibility to my unique gift. As I worked with physicians, I grew to trust what I knew and quieted my self-doubt. I encouraged highly successful individuals to challenge my gift. The accuracy, the genuine insight, after a while gave me the opportunity to settle into me being me with my extraordinary gift of accessing infinite intelligence. After five years of working solely with physicians, I opened a private practice to include private individuals, business entities, and research projects. That was two decades ago, the rest is history.

It has been a fulfilling and thrilling life. I stand firm in my belief that intuition is real, and that accessing and using infinite intelligence changes lives."

For the past twenty-five years, Jennifer has been providing inspirational insight around the globe, contributing in the areas of medicine, science, business, entertainment, and sports. J.K. has worked with licensed professionals from the most prestigious medical facilities in the world.

Jennifer specializes in personal coaching and practical application that can enlighten and support breakthrough growth and transformation for individuals, teams, and business entities. In 2015, Jennifer retired from her one-on-one work. Today she focuses on writing, workshops, and volunteer work associated with her heart.

Jennifer's insight has inspired countless individuals to reach their highest potential with a keen focus and holistic development. She divides her time between the USA, Canada, and France.

BOOKS BY J.K. DICKINSON

INTUITIVE EXPLANATIONS

The Energy of Life System
The Energy of Illness
The Energy of Loss and Grief
The Energy of Parents and Children
The Energy of the Human Spirit
The Energy of Music and Sound Vibration
The Energy of Intuition and Wisdom
The Energy of Religion, Spirituality and Consciousness
The Energy of Love and the Importance
The Energy of Calculated Manipulation

2019–2020
Energy of Life Educational Series
Workshops – Leadership Programs
Fairmont Hotel & Resorts
Four Season Hotel & Resorts
USA – UK – FRANCE – ITALY
JKDICKINSONINTUITION.COM

www.ingramcontent.com/pod-product-compliance
Lightning Source LLC
Chambersburg PA
CBHW040507110526
44587CB00046B/4299